The Organisation of
Science in England

D.S.L. CARDWELL

The Organisation of
Science in England

HEINEMANN
LONDON

Heinemann Educational Books Ltd
LONDON EDINBURGH MELBOURNE AUCKLAND TORONTO
SINGAPORE HONG KONG KUALA LUMPUR
IBADAN NAIROBI JOHANNESBURG
NEW DELHI

ISBN 0 435 54153 6 (Cased Edition)
ISBN 0 435 54154 4 (Paperback)

Published by
Heinemann Educational Books Ltd
48 Charles Street, London W1X 8AH
Printed and bound in Great Britain by
Morrison and Gibb Limited, London and Edinburgh

Contents

List of Diagrams

Preface

This book is the result of my experiences as a practising physicist—a 'pure' scientist—and as an electronics engineer. I had become curious to know why certain topics were included in my particular scientific discipline while others were excluded, or at any rate discussed only cursorily. Generally speaking I wished to know why the various scientific subjects had become specialised in the way that they have. There seemed, on the face of it, no overriding necessity for the specialisation of the sciences, and conventional histories of science threw no light on the problem at all. At the same time I wanted to know how the highly abstract science I had learned as an undergraduate and postgraduate student had been applied to the practical needs of productive industry. When had scientists become professionals, and how had the industrial research laboratory originated seemed to me to be questions of some interest and perhaps even importance.

The recent development of applied science coupled with the industrialisation of science, as J. R. Ravetz aptly terms it, have together brought about what is sometimes referred to as a 'second industrial revolution'. Yet until recently this remarkable development aroused little curiosity. It is probably true to say that the rise of applied science was looked on as no more than the inevitable outcome of science, technology and industrial growth: to be explained, if at all, in terms of some generalised social theory. This was unsatisfactory for it is easy to see that there was nothing inevitable about applied science. It is, *a priori*, the creation of a certain type of society at a certain stage of development. We need to understand how it was related to 'pure' science, established as a consequence of the seventeenth century scientific revolution, to the systematic technology that began in the eighteenth century and, lastly, to cognate social and economic institutions.

One of the first things that must strike us about applied science is that, measured by the time scale of modern history, it is very new. The war between England and France at the beginning of the nineteenth century almost certainly retarded the progress of science by restricting the freedom of most individuals to travel and by impeding

ix

the interchange of books, journals and technical equipment.* But this was incidental. When Sir Humphry Davy and Michael Faraday visited France in 1812 the French Academy saw nothing incongruous in honouring Davy—the most spectacular if not the ablest British scientist of his day—and Davy saw nothing unpatriotic in accepting their honours. Clearly, while science and men of science were held in some esteem, science was not considered to be an activity of profound national importance (at least in England). *Per contra*, under similar political conditions to-day, the able scientist would be the very last man to be allowed to visit an enemy country. Evidently, therefore, the social revolution of applied science was subsequent to 1812 and this is why the period of this study is limited to 1800–1914. The latter date, apart from its obvious implications, can be justified for, as I hope to show, most—though not all—of the factors necessary for the systematic development of applied science had been established in Britain, as well as in Germany and France, before the outbreak of the Great War.

The numbers and status of scientists must always be of overriding importance in determining the speed of a scientific revolution. For this reason I have confined myself largely to a consideration of the men who, at different times, have constituted the body scientific in Britain. It may be objected that the scientist like the poet is born . . . But he is also made in that he is the product of a refined educational system and the potential employee of a large number of specialised organisations. It is to these latter factors that my attention has been directed; but to say more at this stage would be to anticipate later arguments.

At the same time I must enter a caveat. Although I am concerned with universities as scientific institutions, what follows is in no sense a history of 'Oxbridge' and 'Redbrick'; nor is it a history of technical colleges and technical education. Consequently the attention given to various foundations is a function of the roles which, it seems to me, they played in determining the development of 'pure' and applied science. I am incompetent to judge the relative importance of these and other foundations to-day. I have made few references, and those only for comparative purposes, to Scottish and Irish foundations: developments in Scotland and Ireland were significantly different from those in England and Wales and merit independent study.

---

* Up to 1815 French engineers and scientists were largely ignorant of the extraordinary progress made by steam power technology in Britain: a fact that was of some significance for the development of the science of thermodynamics.

During the eighteenth century it was customary for authors to preface their works with fulsome, if not always entirely sincere, tributes to their patrons. Fashions change, and in the nineteenth century occasional references to the authority and omniscience of the Almighty were considered appropriate. Today authors are expected to urge the need for further research and to indicate the directions that inquiries should take. Accordingly I mention a number of problems that I, personally, would like to see investigated. I admit, however, that as a general rule the most fruitful researches are made in the least expected fields.

It was pointed out fairly recently that something like 90 per cent of the scientists who have ever lived are alive to-day. At first sight this estimate looks absurd since the modern scientific community can hardly boast of nine Newtons, nine Faradays, nine Darwins etc, nor can we reasonably expect to witness an advance in knowledge nine times as dramatic as that achieved between the age of Copernicus and that of Einstein. The explanation is, however, quite simple. Within the last hundred years science, formerly the concern of a small band of talented and self-sacrificing devotees, has been transformed into a subject of instruction for vast numbers of students and a remunerative career for suitably qualified graduates. In other words the personnel of science,* the conditions under which it is practised and probably its ultimate prospects have all been radically changed in the last few generations. While there has recently been a gratifying increase in interest in the sociology of science I do not think that the implications of this transformation have been sufficiently clearly grasped. Can we still impute to science the ideals, motives and even perhaps the high prestige that characterised it in the days when it was the affair of a select band of devotees? Can the modern scientist claim to stand in the tradition of Galileo and Newton, Faraday, Joule and Rutherford? From these and similar questions it would seem to follow that what is needed is not so much further research as fruitful discussions between historians of science, philosophers and sociologists on the one hand and scientists on the other. For the present I regard this work as being, in part at least, a contribution towards a social philosophy of science.

Revolutions in the political sense are commonly undertaken to remove injustices and at the same time to bring new social institutions into being. But it has been common experience that revolutions also destroy, along with the bad, much that was good; in some cases the

* The word 'scientist' was coined barely one hundred and thirty years ago.

losses may even outweigh the gains. To the extent that recent developments in applied science warrant the title 'revolution' we may reasonably ask whether they have entailed any losses. It is my belief that there have been losses and I have therefore included, at the end of this work, a brief statement of what I think these losses have been, how they have come about and what difficulties may be involved in trying to recoup them. This particular aspect of the work is necessarily controversial.

In conclusion I turn to the much easier task of acknowledging the help I have received when engaged on the first and second editions of this work. In the preface to the first edition I expressed my gratitude to the late Professors Morris Ginsberg and Douglas McKie for the encouragement they gave me, to the Nuffield Foundation for their support and to the staffs of the various London libraries for their courtesy and consideration. I now want to thank my wife and my colleagues and friends who have helped me to bring out the second edition. In particular I am very grateful to Dr. W. H. Brock, Dr. W. V. Farrar, Professor Donald MacRae, Mr. J. B. Morrell and Dr. J. R. Ravetz; to the postgraduate students in my Department, whose names may be found among the references, and lastly to the library staffs of the Universities of Leeds and Manchester ('Owens' and UMIST).

D. S. L. Cardwell

Manchester, April 1972

CHAPTER ONE

Science
and Society

The nature of natural science is frequently discussed by philosophers and their conclusions are usually given in terms of methodology and epistemology. They are concerned, that is, with the methods that the scientist uses to obtain his results and with the status and significance of the theories and concepts of science in the scheme of rational knowledge. While such investigations are of the greatest importance, especially in view of the specialisation characteristics of modern science, they do not, of themselves, exhaust the question; for science, the development of knowledge and thought, is essentially a social phenomenon [1]. From asking the question 'What is science?' it is a short and natural step to ask 'How did it arise?' or 'Under what circumstances can we expect it to be actively pursued?' and it is at once apparent that, since scientists are not logical machines operating independently of their environment, these questions involve social factors. Indeed, we can say without further consideration that science is a variable of society and a very complicated one too; for, while many different societies have evolved advanced systems of law and philosophy and refined forms of art, only one society—our own—has possessed those vital elements that made possible the systematic and widespread development of the advanced sciences, and has succeeded, moreover, in utilising science to solve problems in industry and the arts.

The grandchild of ancient learning, this uniquely successful method of interpreting nature, was born during the Middle Ages; a time of intellectual ferment and of great mechanical inventiveness. Aided by initial contributions from the Arabs and Indians, science advanced slowly and painfully to the time when the masterly Galileo set the pattern for its full development. During and after this period it was in intimate relationship with the current philosophical and religious ideas, with the advance of technics and with the practices of the fine arts and medicine; drawing its inspirations and conceptions quite impartially from all these distinct activities and, in return, making its contributions, material and intellectual, to the development of society. To elaborate this theme would be, in itself, to attempt a work of major scholarship; but one point can, I think, be made: to talk of the 'impact of science on society, is to oversimplify the issue. Science is not an alien, external force like famine, pestilence or conquest; it is a characteristic of our society. It is made by men in that society and the relationship between their work and the social whole is both subtle and complex.

In the most general terms we can presume that the successful prosecution of science depends upon a number of internal factors, the chief among which are: the cultural heritage of abstract knowledge and practical techniques, the free circulation of ideas and constructive criticism, freedom of research, the adequate endowment of research and the state of the ancillary educational system. We should, therefore, expect to find that at all times, past and present, the predominantly important social institutions will play their parts in determining the conditions of the internal factors and hence the level of scientific activity. In particular, religious and philosophical ideas, technics and economic institutions may all have a bearing on science although their effects will not, of course, be uniform and of equal importance in different periods. Thus we may expect to find that the state of science will be correlated with the social institutions of different peoples at different stages of development. But here a peculiarity of science should be noticed: it is, of all human activities, the most truly international, for it is only in a limited sense that we can speak of 'English science'.

It is true that we can, without much difficulty, recognise English, French, German styles of science, each one reflecting, no doubt, aspects of national character [2]; but these 'styles' are, in the last resort, superficial, it is the factors that are common to all that are important. When we speak of 'English law', 'French literature', 'Italian art' etc. we may well be noting, indeed emphasising, the significant differences from the practices of other countries and peoples. But the language in which a scientific memoir is written is of no importance—it loses nothing in translation. Therefore, to a greater extent than in the cases of other intellectual activities, the scientific achievements of any one country are a function of those of its neighbours. While this view rejects scientific chauvinism of the type: 'Chemistry is a French science. It was created by Lavoisier...' (Wurtz), it does not prevent a reasoned use of comparative methods.

Science, then, is a European endeavour and any claim by one people, or nation, to pre-eminence or to be the founders of any particular science cannot be accepted. This odyssey of the European mind has survived the difficulties and dangers of the profound social, political and religious changes of the last four centuries. On the other hand, although we have a most inadequate understanding of the factors which favour the development of science in a given society, it is not overbold to maintain that, had the terms of trade changed against science at any time during that period, the scientific enterprise would have been snuffed out as effectually and finally as the Divine Right of Kings or the Rule of the Saints.

If these preliminary generalisations are accepted, as I think they must be, we are faced with the question as to which is the most profitable way of studying the social aspects of science. Here, the field is so wide and the possibilities so varied that choice is almost entirely a personal matter—the relations between scientific achievement, in the material sense, and scientific thought, in the philosophical, with other leading social activities are so profound and many-sided. However it is reasonable to argue that as science is made by men in society the best way is to begin, continue and end with that group of men who practise science and to try to elucidate the factors that have either advanced or retarded their work. This seems to me preferable to attempting to study the variation of scientific activity with economic changes or with the development of philosophical ideas, for until we have some clear idea as to what the scientist is, as a man in society rather than as an impersonal producer of information, such studies may lead to unwarranted correlations.

A BRIEF CONSIDERATION OF SOME THEORIES

To develop this point of view and, at the same time, to illustrate the difficulties of the theme we shall briefly consider certain theories put forward to account for the varying levels of scientific activity at different times and in different societies.

In his stimulating little volume on Greek science, Professor Farrington asserts that ancient science perished through the social effects of slavery which, by degrading labour and technics, necessarily cut science off from one of its most important sources of inspiration. It should be remembered, however, that the theory is not original. For example, it was explicitly propounded by Justus von Liebig as long ago as 1867 [3] and, in 1870, Lyon Playfair in referring to Greek science, remarked:

'A citizen with slaves crushed invention lest it should interfere with their value on the market . . .' [4]. A popular writer of the same period, Winwood Reade, made much the same observation in his book, *The Martyrdom of Man*.* The opposite point of view has recently been cogently argued by Bernard Barber who points out that Greek science had a long career in an epoch when slavery was always a dominant social institution [5], and another modern writer doubts whether '. . . even in classical antiquity the separation of technics and science was as complete as has sometimes been supposed' [6].

It seems likely that ancient science was a complex affair and that even Greek science did not have the unity that we usually ascribe to it. The Greeks were a mercantile, seafaring and colonising people; they were not greatly interested in technics and their science tended to be abstract and speculative. This was an obvious defect but an opportunity to remedy it came when the Greek and Egyptian civilisations came into intimate contact after the foundation of the great city of Alexandria. The Egyptians were more concerned with technics and had surprisingly few inhibitions about nature; for example, they allowed the dissection of human corpses [7]. Accordingly there was, for some time, a promise of a genuine experimental and theoretical science developing at Alexandria. Here, Eratosthenes had measured the radius of the earth, Hero and the members of his school studied the problems of gearing and may even have experimented with model steam engines, Ptolemy brought ancient astronomy to its highest point and, by tradition, experimental chemistry was born. Perhaps there was some significance in the fact that the lighthouse at Alexandria was the only one of the seven wonders of the ancient world that was useful and not an instance of conspicuous consumption. But in the end nothing came of it. Probably because, as we now know, if science is to flourish more than one centre of learning and research is necessary and Alexandria had no rival in the ancient world: possibly, too, because slavery continued to poison the atmosphere and retard the development of science (or perhaps we should call it proto-science?).

Although it does not strain credulity to believe that a society which could regard many of its members as 'living tools' could not provide effective stimuli for technics and science, it would nevertheless be unreasonable to suppose that the institution of slavery is alone sufficient to account for the decline of ancient science. If we suspect that a uni-causal theory is an oversimplification, if we admit that other factors may

* Watts & Co., London, 1932 edition, pp. 404–5.

well have been effective in determining the level of scientific activity, we must at once concede that the reverse proposition is, at least, feasible: that inability to forward science and technics contributed to the protracted existence of the obnoxious and inhibiting social institution of slavery.

Whatever factors led to the collapse of ancient science, or proto-science, there can be no doubt that, in the freer atmosphere of medieval Europe technics flourished as they had never done before. From the reign of Charlemagne onwards the Europeans showed themselves willing to accept inventions from whatever outside source they came. As time went on classes of artisan/inventors and artist/inventors emerged: men who were capable of making radical inventions on their own account. Thus paper and gunpowder were introduced, the mariner's compass appeared and spectacles, firearms, the weight-driven clock, the windmill and the spinning wheel were invented; at the very end of the medieval period came Johan Gutenberg's revolutionary printing press.* With this inventiveness and willingness to innovate there went a deep interest in machinery and a boldness in the face of nature that surpassed any other civilisation. The importance of this for the establishment of science—the 'mechanical philosophy'—in the seventeenth century does not need to be stressed.

Among the additional factors which we presumed might govern the development of science were the leading religious, philosophical and cosmological beliefs current in society. This suggests that it was no accident that the rebirth of science, together with the recovery of ancient learning in the twelfth century, occurred at a period of history well characterised as the Age of Faith. The significance of this coincidence has been commented on by Whitehead who, in a famous passage, related the rebirth of science to medieval insistence on a rational Deity and on a correlative rational Creation; it being argued that, without prior conviction that the universe is rationally ordered and susceptible, therefore, of rational interpretation, the scientist would never commence laborious and difficult researches.

It is, for example, undeniable that, after the close of the medieval period, Johann Kepler and, from a different philo-sophical stand-point, Galileo [8] were both convinced that nature is basically mathematical and that its laws could be intuited by mathematicians. Once Galileo had demonstrated the validity of this belief subsequent researches and studies, culminating in

* The mechanical clock and the printing press still condition our lives to an extent equalled by very few other inventions.

Newton's *Principia*, could only confirm it beyond reasonable doubt.

On the other hand the universal acceptance of Christianity in western Europe implied—from the scientific point of view—more than an adjustment of philosophies; it entailed the emergence of entirely new social organisations. The Graeco-Roman world had been widely tolerant of different religions and different philosophies, ranging from the 'scientific' determinism of the atomists to various kinds of mysticism. It is difficult to see how such institutions as universities could have come into being under these circumstances; for universities imply some generally accepted corpus of knowledge, some basic philosophy, that it is their duty to augment and to transmit to the following generations. As opposed to transient and informal schools of philosophers the development of universities seems to be an almost exclusively European achievement. The learned monastic orders, Franciscans and Dominicans, the open universities—'cosmopolite corporations' as Sir William Hamilton called them—with their Regent Doctors, Masters of Arts and the scholars wandering from teacher to teacher, from university to university, were all conspicuous features of the medieval scene. Coupled with developing technics the universities could hardly avoid exerting a widespread influence as centres for the diffusion of new ideas of all sorts irrespective of the inner content of the established religion and the associated metaphysic.

Another study of the relationship between religion and science, this time in seventeenth-century England, forms a major part of a famous paper by R. K. Merton [9]. Following a suggestion of Max Weber, Merton examined in great detail the connections between Puritanism and science; and after careful analysis reached the conclusion that the Calvinist cosmology coupled with the characteristic 'ethic' expounded in sermons and religious writings, provided strong stimuli for the pursuit of science. This, he claimed, was validated by the large number of Puritans active in seventeenth-century science and associated with such enterprises as the foundation of the Royal Society. It is important to notice that Merton does not maintain that religion is the independent and science the dependent variable; nor does he suggest that a set of religious beliefs is sufficient to account for the emergence of great scientists, the Newtons, Hookes, Halleys and Boyles of this world. On the contrary he says quite explicitly that the relationship between religion and science was one of reciprocal reaction.

The Puritan, without intervening authorities to interpret and mediate, was directly accountable to God for his career on earth: an earth that was, together with the rest of the cosmos, God's great Creation and whose study must, therefore, be in itself an act of reverence. Calvinism thus provided strong grounds, psychological, metaphysical and social, for the pursuit of science. It is admitted that the theory is of limited application: the role of Calvinism in inspiring scientific work should not be generalised. Scientific academies, which are accurate indices of public interest in science, were founded in many countries—Lutheran, Calvinist and Catholic—contemporaneously with the Royal Society; and the magnificent French achievements in science in the seventeenth and eighteenth centuries owed nothing to Calvinism.

Recently P. M. Rattansi has shown, convincingly, in two important papers [10], that the science of mid-seventeenth-century England and of the early Royal Society was more complex than we had previously supposed. The 'mechanical philosophy', which was the viable form of science, was accepted by only a section of the scientific community. A significant part of this section, which included Hooke, Hobbes, Wren and, for all practical purposes, Newton, tended to be Anglican in matters of faith and conservative, or Royalist, in politics. Many of the radicals and Puritans, on the other hand, favoured the magical or anti-rational 'sciences' of Paracelsus and van Helmont; under the Commonwealth there was even a movement to establish the teaching of these 'sciences' at Oxford and Cambridge. It seems likely, therefore, that the 'science' that many Puritans favoured was not of a kind that we would gladly recognise to-day.

On the need for further comparative studies of religion and science, Merton mentions the suggestion that Protestantism is generally more favourable towards science than is Catholicism. In support of this thesis, Protestant Scotland with her long line of famous scientists, has been compared with Catholic Ireland who can claim only one: John Tyndall. The Irish, Merton comments, make up for their lack of science by excelling in the histrionic arts, which are disapproved by Calvinists. But an Irishman could reply, very fairly, that Calvinist Wales has yet to produce one world-famous scientist while it is undeniable that the Welsh are successful practitioners of the histrionic arts, however repugnant they are supposed to be to Calvinists.

B. Kelham, who has studied the history of science and scientific institutions in Ireland [11] has concluded that, all things considered, the surprising thing is not that there was so little science in that country, but that there was so much. Earlier studies, by Galton and by Havelock Ellis, had tended to be biased in that a man born in Ireland who made a scientific reputation in England or Scotland (e.g. Lord Kelvin) was classified as English or Scottish while an Englishman or a Scotsman who did scientific work in Ireland was still classified as English or Scottish. More serious has been the omission of those Irishmen who emigrated and made names for themselves as scientists or engineers in the Dominions and, especially, in the United States. The numbers of such men increased greatly after the disastrous famines in Ireland during the 1840s and an impressive list of names can, in fact, be drawn up. It is true that the Catholic Hierarchy opposed or, at least, snubbed, the scientific institutions, including the university colleges, that the London government established in Ireland during the nineteenth century. This was unwise but the Hierarchy could have argued that their opposition was not to science and scientific education as such but to the attempts of an heretical, even alien, government to impose its ideas on Catholic Ireland. Parenthetically, we wonder whether English and Scottish Calvinists would ever have been enthusiastic supporters of science had England and Scotland been Catholic countries and the Royal Society been a predominantly Catholic body? The conclusion we are left with is that the Mertonian thesis, while interesting and important, is unproven.

Other social institutions: industry and commerce have clearly influenced the rise of science since 1600. The relationships between mining and the metallurgical arts on the one hand, and mechanics and chemistry on the other, are obvious *a priori*, and are abundantly confirmed by the facts of history. The industrial and economic factors in the growth of science, indeed the industrial origins of science and the correlative importance of the artisan and technician in contributing to scientific progress, were repeatedly stressed by Victorian writers like Whewell, Playfair, Lockyer and others. That theories of the social motivation of science should become widespread in a utilitarian age is not, perhaps, surprising; especially in a society becoming increasingly aware of the industrial and social importance of science and, at the same time, moving towards the liberal democratic State. Accordingly, Playfair was voicing a fairly general opinion when, over a hundred years ago, he asserted that the springs of action in science were located in the industrious classes, while, on the other hand, an aristocratic society was not—and never had been—one which would be favourable for the advance of science.*

* The example of pre-Revolutionary France refutes the second part of this generalisation.

There is some danger that theories of this nature become unduly extrapolated and the advance of science reduced to the status of a dependent variable of what are alleged to be large-scale social or economic movements. In these cases science is regarded mainly, or entirely, from a utilitarian point of view; it is held to be the 'cutting edge' of engineering and technology and, as such, entirely subordinate to, and dependent on, social needs. In brief, it is said that science is no more than a means of extending mastery over nature. Sometimes a moral imperative is added: Comte, it will be remembered, applied the test of fecundity to science and would, as a result, have prohibited such 'useless' studies as astrophysics. But criteria of this kind cannot be used: on what conceivable grounds can we justify the assumption that a science that is useless to-day will be useless to-morrow? Is it not possible that it might prove to be of the greatest practical benefit? Science deals with the at-present unknown; how can current social needs determine which of a number of unknowns will prove to be of the greatest fruitfulness in the future?

Let us consider, as an example of the social determination of science, the proposition that the advance of science, either generally or during certain historical periods, can be understood in terms of the economic and technical requirements of certain social classes. If major scientific developments take place in response to socio-economic needs, then, *ex hypothesi*, an examination of the history of science should reveal a pattern of development of the various sciences that can be interpreted accurately by reference to the economic history of the appropriate social classes. Even if we grant for the sake of argument that such parallels can, in fact, be found, before the interpretation can be valid it is necessary to *prove* that the problems in every science, at any given time, are all equally capable of solution, requiring only a biased social stimulus for one group of problems, rather than another, to be solved. In other words we must be quite sure that the agent governing scientific development is the external, social, one. Specifically this means that we must prove that, in the seventeenth century for example, chemical, biological, geological etc., revolutions were all possible, equally with the Newtonian revolution, and could, therefore, have taken place had the appropriate social stimulus occurred. It is, however, impossible to see how this proposition can be proved; yet without it we cannot even begin to justify the above interpretation, and the theory falls to the ground. And, for that matter, any theory that would explain the advances of science in terms of generalised social institutions must either fail or be tautological, despite the plausibility with which correlations of this kind can sometimes be established. In fact, we do not usually require theoretical analysis in order to refute such theories; we find, for example, that the doctrine of utility breaks down when we attempt to apply it to such important sciences as electricity and sidereal astronomy, both of which were pursued with zeal and ability for several hundred years while remaining of little or no practical use. To take another, and very familiar, example, it has been said that the expansion of sea-borne trade was the great incentive to that work which culminated in Newton's *Principia*. An incentive no doubt it was, but it is very far from clear just what interests Copernicus, Kepler, Descartes and Galileo had in maritime affairs, or, indeed, what common economic motive can be detected in their works.*

The great advances in mathematics, mechanics and astronomy from the time of Copernicus onwards owed much to fruitful cross-fertilisation between these sciences as well as to certain technical innovations such as the invention and development of the telescope. The process of discovery and inference culminated in the realisation that the solar system could be regarded as a working physical model, the details and movements of which could be accurately described. What was required was an interpretation and this Newton was able to provide. On the other hand the non-Newtonian sciences, while not without many devotees of genius, were not able to make such revolutionary progress: in chemistry, biology etc., the phenomena were complex and there did not exist any form of 'working model'. The great correlations in these sciences were yet to come.

This brief argument suggests that, apart from those social institutions which, at different times, provide mental and material incentives for the pursuit of science, two further factors governed scientific development. These were, firstly, the existing structure of knowledge together with what may be called the inherent opportunities of the situation and, secondly, the quite unpredictable emergence of men of genius and their response, in terms of rational creative thought, to the opportunities presented.

While these, and possibly other, factors determined the level of scientific activity there was virtually no formal organisation of science: there were no structures of authority and subordination, no limitations of function, no professional associations etc. Science was conducted by

* Sir G. N. Clarke dealt with this aspect in his *Science and Social Welfare in the Age of Newton* (Oxford University Press, 1937).

amateurs and even research of admitted national importance was left to individual goodwill: John Flamsteed, the first Astronomer Royal, had to equip the observatory at his own expense.

Even the universities, the patrons of learning, failed to achieve any notable organisation of science. In the early days university education had been based on the study of the seven liberal arts followed, after graduation, by continued study in the main faculties of divinity, law and medicine. The liberal arts and one of the professional faculties were, judged by contemporary standards, of a scientific nature; a fact reflected in the men of science to be found at medieval universities: early Oxford had a famous school of astronomers besides individual pioneers like Robert Grosseteste. Unfortunately these institutions did not prove sufficiently adaptable and, in later years, rigidity crept in; studies became formal, philosophies dogmatic and unalterable. The reaction, when it came, took the form of a protest by the humanists against what had become the dry pedantry of the scholastics and the reform, which the humanists achieved, was the incorporation of the study of classical literature in the arts faculty. This form of liberal education became increasingly important in the universities and professional training was gradually abandoned to such extra-mural institutions as the Inns of Court and the London hospitals.

The decline of endowment of educational foundations during the seventeenth and eighteenth centuries coupled with the triumph of Aristotelean educational theory at the universities indicates that university education was becoming increasingly the prerogative of the upper classes: a literary education is an upper class education. This was hardly a favourable conjunction for university science.

During the years of decline, therefore, the brilliant assembly of scientists at the universities over the period of the Newtonian revolution stand out as a defiance of circumstance. They did not mark the beginning of the modern universities, conceived as centres of research and active scholarship; rather they represented the last and greatest efflorescence of the medieval schools. (Was it not symbolic that Newton himself entered college as a sizar: as a 'poor scholar'?)

REFERENCES

[1] The most perceptive discussion of the significance of social factors in the early development of science is Hans Kelsen's *Society and Nature: a Sociological Inquiry* (Kegan Paul, London 1946). See

also L. T. Hobhouse's article 'Evolution and Progress' in *Social Evolution* (Columbia 1911).

[2] On the problems of 'national character' see Morris Ginsberg, *Reason and Unreason in Society* (Heinemann, London 1948), pp. 131–55. An interesting example of a highly prejudiced view of national character in the field of science is provided by Pierre Duhem in *Aim and Structure of Physical Theory* (Princeton University Press, 1954), p. 55 et seq.

[3] Justus von Liebig, 'The Development of Science Among the Nations', a lecture delivered in 1867 (Edinburgh 1867).

[4] Lyon Playfair, 'The Inosculation of the Arts and Sciences', an address given in Birmingham, September 1870. Printed in *Subjects of Social Welfare* (London 1889).

[5] Bernard Barber, *Science and the Social Order* (Allen and Unwin, London 1953).

[6] A. C. Crombie, *Augustine to Galileo* (Heinemann, London 1961) i, p. 175.

[7] For a discussion of the various tabus on scientific inquiry in the early days see Robert Lenoble, 'La Pensée Scientifique' in *Histoire de la science*, edited by Maurice Daumas (Editions Gallimard, Paris 1963).

[8] Alexandre Koyré, *Etudes galiléennes* (Hermann et Cie., Paris 1939).

[9] R. K. Merton, 'Science, Technology and Society in 17th Century England', *Osiris* (1938) iv, p. 360–632; *Social Theory and Social Structure* (The Free Press, New York) pp. 628–49. See also Max Weber, *The Protestant Ethic and the Spirit of Capitalism* (Allen and Unwin, London 1930) Ch. 5, note 145, and the essay by A. R. Hall, 'Merton Revisited', *History of Science* (1963) ii, pp. 1–16.

[10] P. M. Rattansi, 'Paracelsus and the Puritan Revolution', *Ambix* (February 1963) xi, pp. 24–32. 'Politics and Natural Philosophy in Civil War England', *Actes du XIe Congrès International d'Histoire des Sciences*, ii (Warsaw 1967) pp. 162–6.

[11] B. B. Kelham, *Science Education in Scotland and Ireland, 1750–1900*, Ph.D. thesis, Manchester University (1968).

CHAPTER TWO

Eighteenth-Century
Background

Although the scientific achievements of the eighteenth century were substantial, especially in France, the technological triumphs were even more important. For one thing there was that unique event, the industrial revolution, which began in England in the second half of the century; for another there was the infusion into the technical arts of the scientific ideas worked out in the seventeenth century so that *technology*, or technics based on science became possible. It was, for example, only after the establishment of Galilean mechanics that it became possible to express the power of an engine in quantified terms —a weight lifted a unit distance in a unit time—and the efficiency of an engine as the ratio of the work done to the effort put in. Before the science of mechanics was established power and efficiency could only be expressed in qualitative measures: a machine was a good one if it was made of sound materials, was reliable and could cope with all the requirements put upon it. After the seventeenth century, then, the assessment of machines became quantitative [1] and the measures of power and efficiency were established.

With the rise of technology in the eighteenth century the different modes of innovation became clearer and more complex. Francis Bacon had, in the seventeenth century, recognised two different forms of innovation; the eighteenth century was to add two more.

In the first place, according to Bacon, there are inventions that are quite independent of contemporary knowledge and that could have been made at any time or, at any rate, could have been immediately understood by (let us say) Archimedes. We shall call these empirical inventions and among them we include the medieval spinning wheel, Arkwright's automatic spinning machine (the 'water frame'), the turret lathe, and the zip fastener. Empirical inventions may be complete in themselves, like the ones just mentioned, or they may form parts of more complex innovations, in the way that Watt's invention of the parallel motion was an important part of his rotative steam engine.

In contrast to empirical inventions are those that can only be made

13

if appropriate knowledge is available. Thus firearms and the mariner's compass would have mystified Archimedes since he knew nothing of gunpowder or of the north-seeking properties of the lodestone. To-day we call such inventions 'science based' and among them we would include the Newcomen engine of 1712 (which worked by harnessing the newly-discovered pressure of the atmosphere), the electric telegraph, radio, radar, television and the atomic bomb. These inventions have, as Bacon predicted, become much more numerous and important. Again, these inventions may be complete in themselves or components of more complex innovations.

A distinctive contribution of the eighteenth century was the technique of systematic improvement. This appears to have begun, at least as far as England was concerned, with the work of John Smeaton (1724–92) and may owe something to Newton's experimental methods (*Opticks* 1704). All the components and processes of an engine or machine are kept constant save one which is varied systematically. The particular variation that results in the optimum output is noted and the procedure is repeated for each component and process in turn, so that ultimately, the best possible performance can be obtained. Commonsense as this procedure may seem to be, it is quite clear that it could hardly have been thought of before the post-Galilean concept of quantifiable efficiency had been elucidated. It is therefore a sophisticated mode of innovation but it is important to realise that it cannot, of itself, lead to radical new inventions, only to evolutionary improvement.

The second contribution of the eighteenth century was the first tentative indication of applied science. But before going further it would be as well to define what we mean by science so that the chances of ambiguity are reduced. We may define science as an endeavour to make sense of the external world by the progressive establishment of universal laws of nature. Such laws have hypothetical status in that they may be superseded, modified or subsumed within more general laws. In fact, the inclusion of particular laws within a more general law is a hall-mark of the progress of science. The testing of scientific laws and of the validity of the concepts in terms of which they are expressed, is carried out by means of recognised experimental techniques. Since the seventeenth century occult and magical elements have been banished from nature and the physical sciences with their distinctive 'mechanical philosophy', which was established in that century, have been particularly successful in reducing a wide range of physical and chemical phenomena to scientific law of mathematical and mechanical form. Lastly, we should note that, as J. R. Ravetz has recently shown, scientific work is essentially a developed and highly skilled craft, evolved over centuries of experience, and using its own distinctive 'tools', experimental and theoretical.*

The term 'applied science' can be, and often is, used to describe the practical application of the laws of science. A technologist or engineer building a new machine, plant or structure will base his designs on the established laws governing the materials and processes involved under the conditions to which he knows they will be subjected. Although this may well involve complex calculations and will require detailed knowledge of scientific laws and principles, the technologist does not envisage testing these laws, nor does he seek to discover more general ones. Indeed in such cases the laws are accepted as given and their validity is unquestioned. Evidently, should the relevant laws be invalid then disaster may result. Application of the laws of science may therefore be important economically and socially, but does not generally involve innovation as regards ideas or techniques.

The proper use of the expression 'applied science' should be, I think, to describe the actual investigation, by the methods of science, of the processes or products relevant to the 'hardware' of particular industries. It is a science restricted to the foreseeable interests of industry. Applied science, so defined, can be sub-divided into two categories. One type is concerned with the actual processes of industry, with the efficiency of operation of plant or machine. Good examples of this type of applied science were Smeaton's investigations into the relative efficiencies of water-wheels and Liebig and Playfair's researches into the operation of blast-furnaces. The other type is concerned with the products of industry, and takes the form of a search for new materials or articles for public consumption, as exemplified in the modern world by the systematic discovery of new pharmaceutical products, new plastics, improved transistors, etc.

On the social scene the widespread practice of applied science is manifested by the industrial or government research laboratory and by the existence of a distinct group of professional applied scientists. (This requires no proof. It is an evident fact of the modern social scene.) In spite of its importance, this activity, as we have already pointed out, is very new; in England it has been practised for no more than eighty years. It is obvious that the use of such sophisticated

* J. R. Ravetz, *Scientific Knowledge and its Social Problems* (Clarendon Press, Oxford 1971) p. 75 et seq.

techniques as a matter of industrial practice represents a revolutionary change. But it is not at all clear that applied science, with its research laboratories and their staffs of professional scientists, should have developed smoothly, logically, from the common practices of productive industry. The question therefore arises—how did this come about? And it is at once apparent that it is a very complex and difficult question. It involves the changing economic structure of the country, the advance of science and a whole array of social questions, each of great difficulty. We cannot hope, in what follows, to do more than examine a few facets of the problem.

It is true, *a priori*, that for applied science to be practised on any scale there must exist in the given society a general appreciation of the value of science and a knowledge of the possibilities of its application. There must also exist a recognisable class, or group, of trained professional scientists financially dependent on their vocation for their livelihoods, and this class, in turn, implies the existence of an educational system properly equipped to provide that training. Necessary, too, are research laboratories, maintained by the State or by industry or by both, wherein the vocation can be practised. Underlying, and determining, these essentials are the conditions of the society and the received ideas of public welfare. Judged by these criteria it is clear that applied science could only be practised on an individualist basis during the eighteenth century. Among the first, and certainly one of the most important, of the early applied scientists was James Watt. His researches into the distribution of heat during the operation of a Newcomen engine, researches on which he based his invention of the condensing steam engine, form a paradigm of applied science [2]. But the particular circumstances and the abilities of the man himself were both exceptional.

There remains one more condition to be fulfilled before applied science can be substantially practised. Evidently, before you can apply science you must, in the manner of Mrs. Beeton, have a science to apply; and over the greater part of the eighteenth century the only science to have achieved a degree of maturity was that of mechanics. Chemistry, the biological sciences and a good deal of what we now call physics were at a 'progressive' empirical stage. The different branches of these sciences were independent of each other; they had little or no theory, very few tested laws and a multiplicity of concepts. It is very difficult to see how systematic applied science could be practised in these circumstances. How, for example, could there be a scientific 'applied chemistry' when the phlogiston theory dominated that science?

Despite the individualism of the different branches of science and their consequent isolation from each other, the eighteenth-century scientist was not, generally, a specialist. His contributions were not limited to one branch of study but often ranged over very wide fields indeed. For example, three of the most notable scientists of the time—Black, Priestley and Cavendish—made contributions to physics* as well as to chemistry. Moreover, it was very probable that an eighteenth-century scientist would be an amateur. That is to say he would not be gainfully employed as a scientist, or, if self-employed, there would be no evident connection between his occupation and his science. The word 'amateur' as used in this context should not be equated with 'dilettante', still less does it imply an inferior degree of ability. Perhaps this will be clear if I point out that such amateurism was not confined to the natural sciences: Gibbon, for example, was an amateur historian and Hume an amateur philosopher. Men like these, whether scientists or not, are best described as devotees, and the quality of their amateurism can be properly inferred from the tributes paid to them. Thus: 'His leisure and his fortune were devoted to the promotion of science' (Rev. John Michell) [3], and: 'Geology had kept him poor all his life by consuming his professional gains' (William Smith) [4].

This raises the question of the extent to which amateurism was characteristic of science at this time. Before I give an estimate of the number, it must be stressed that it is extremely difficult to separate science from its collateral activities; 'pure' science shades off imperceptibly into applied science; also, in another direction, into philosophy. We can, no doubt, separate science from its near relatives—engineering, technology, philosophy—using the criteria of the methodologist. But when we consider the personnel of science the problem becomes far more complex—the same man might be a scientist one day, a technologist the next. If we limit ourselves to those scientists who contributed significantly to the advancement of rational knowledge, irrespective of whether their motives were technological or philosophical, we find that of 106 leading scientists of the century, some 40–50 must be classified as amateurs, or devotees, while in many of the remainder the amateur element was strong.

* This is an anachronistic use of the word. The term 'physics' did not become common until the later nineteenth century. Prior to this, 'physics' was partly subsumed under 'natural philosophy'.

A significant contribution can only be inadequately defined as one that was of fundamental importance and proved fruitful for later development. Obviously it is impossible in the space available to justify the selection of each one of the 106 scientists; their names, and thus the reasons for their selection are to be found in the sources for the history of science as well as in such reference works as that of Ludwig Darmstaedter [5]. In most cases biographical details can be obtained from the *Dictionary of National Biography* and other sources, although in a few cases too little is known to permit of classification.

It is even less easy to draw a line between amateur and professional. In a number of cases there was an oblique connection between the profession and the science. Consider, for example, the case of a medical man disposed to research. If he made contributions to anatomy and physiology we must classify him as a professional scientist: there was a connection between his profession and his science. If, however, his tastes ran to botany or to astronomy it is correct to describe him as an amateur. Again, some scientists started as amateurs and ended as professionals, as did Sir William Herschel the great astronomer who was originally a musician. Others started as professionals and ended as amateurs, as did several Cambridge men who resigned their Fellowships on being awarded Church Livings. Clearly, in these border-line examples, classification must be governed by judgment of the individual cases and no hard and fast criterion can be set up.

The three leading professions represented among the 106 scientists are medicine, technology and the church in the percentages of 20, 15 and 10 per cent respectively. (These figures are, of course, very approximate. Technology includes such occupations as engineering, surveying, and instrument making.) The third learned profession is not well represented and, in fact, lawyers—with some very distinguished exceptions—do not seem to have shown quite the same taste for science as the clergy and, *a fortiori*, doctors have done. Possibly this was because the pursuit of law has always been a highly specialised occupation, demanding the closest attention from its votaries if the greatest rewards, intellectual and material, are to be achieved. Also, of course, the study of law has never had any direct relationship with the pursuit of natural science.

Corresponding to this prevailing amateurism was the state of the English universities which, continuing in decline, partook in liberal measure of eighteenth-century indolence. An historian of Oxford University, A. D. Godley, tells us that they were denationalised; being monopolised by the upper and middle classes while the yeoman class—Newton's class—was but scantily represented [6]. In this they compared unfavourably with the Scottish universities which were far more democratic and 'open'. Moreover, the application of the Tests excluded from the universities a large proportion of the population: dissenters, Roman Catholics and Jews. Within the universities election to office was too often the result of favouritism and a Fellowship or Chair, once achieved, was too often regarded as a sinecure; the obligation to lecture or to teach being conveniently forgotten. The studies of law and medicine remained outside the province of the universities and the greater part of those students who were not to lead a life of leisure read for the B.A. degree; most of these probably intended to take Orders. The universities were dominated by the wealthy and indolent colleges to such an extent that the Oxford and Cambridge of those days could be described, not unfairly, as groups of exclusive, if somewhat eccentric, clubs. Consequently the number of matriculants fell steadily, not only relatively but absolutely, so that by the end of the century the nobility and the very wealthy formed a disproportionately large element in the undergraduate population. This could hardly pass without comment, even in a very indulgent age. In 1776 Adam Smith submitted the universities to the test of utility: 'Have those endowments', he asked, 'contributed in general to promote the end of their institution . . .? Have they directed the course of education towards objects more useful both to the individual and to the public, than those to which it would naturally have gone of its own accord?' He answered his own question: 'In the universities the youth neither are taught, nor always can find any proper means of being taught, the sciences which it is the business of those incorporated bodies to teach' [7].

On behalf of the universities it is only fair to remember that they were, in large measure, the victims of obsolete but immutable statutes imposed on them by Tudor and Stuart governments. Furthermore in an age when nepotism and its camp-follower incompetence were common to all organisations it would have been remarkable had the universities escaped the contemporary failing. Yet, nothwithstanding the observations of men like Adam Smith, Gibbon* and others, there must always have been a sound core of hardworking students and many of these must have come from homes that were by no means wealthy. Also there were, among Fellows and Professors, some able men of science. Roger Cotes, Richard Watson, William Whiston, George

* *Autobiography* (Everyman edition, London 1948) pp. 37–50.

Atwood, Isaac Milner, William Farish, Edward Waring and others carried on the traditions of learning and research and, in some cases, gave successful lectures on experimental science and technology.

The most interesting university developments of the time were the rise of mathematical studies and the gradual emergence of the examination system, both of which took place at Cambridge and both of which indicated, and implied, reform.* The importance attached to mathematical studies owed something to the fame of Newton but it was also valued highly as a part of a liberal education; in which respect it probably owes much to the traditions of English Platonism [8]. On a less philosophical plane, mathematics was regarded as a suitable mental discipline for those who intended to take up the profession of law. But there was no idea that the training of professional mathematicians and 'physicists' was a desirable object; nor was it generally considered that the universities should contribute to the advancement of mathematical science. In the light of this attitude it is easy to understand why there was no associated school of progressive mathematicians. An adherence to Newton's strict geometrical forms and to his fluxional notations meant that the new analysis and rational dynamics as developed in France were neither taught nor understood in England. Not until 1803 was a book on analytical methods published in this country.

Meanwhile the old oral disputations—'acts and opponencies'—were being gradually supplemented by, and were eventually to be superseded by, written examinations with the questions at first orally delivered but later given on printed papers. The justification for the written examination was that it constituted a fair test for those aspiring to honours or to college office. It tended to reduce favouritism and was, moreover, efficient and economical in separating the sheep from the goats, the able from the obtuse. Before the century was over, Dr. John Jebb and his friends were trying hard to make it compulsory for all Cambridge students to submit to examination once a year. In this they failed, but the attempt itself was indicative of changes to come [9].

The dissenting community, excluded from the universities by repressive legislation, were faced with the problem of the higher education of their youth and the training of their clergy. Wealthy dissenters could, and frequently did, send their sons to Scottish or Dutch universities, but such a course was open only to a few. It was to meet a more general need that the well-known 'academies' were established.

* The first Mathematics Tripos List was in 1747.

At first fugitive and peripatetic, these colleges were always served by devoted and capable men, and some establishments, notably the one at Warrington, later achieved great distinction in staff, students and standards of work. Towards the end of the century, with the spread of Unitarianism, there went, *pari passu*, a passionate zeal for intellectual freedom and a keen interest in natural science; previously the emphasis had been rather on the classics and theology, although science, together with law and philosophy, had never been neglected [10]. Some of these academies were even able to form collections of 'philosophical apparatus'—air-pumps, frictional electric machines and the like—for demonstrations to the students.

But the colleges were unendowed and, with no State aid, their continued existence could not be other than precarious. Although a few managed to survive, eventually to become theological colleges or schools the greater number perished before the end of the century. Strangely enough their demise was expedited by the repeal of the laws that prohibited dissenters from teaching.

A second factor related to the prevailing amateurism was the undeveloped state of many branches of science. Not until 1735 did the first chemistry textbook, Boerhaave's *Elements of Chemistry* [11] appear in English, and books on applied chemistry did not become common until the end of the century.* The decisive change in this respect occurred when Lavoisier established scientific chemistry; this gave chemical manufacturers a much better understanding of processes and products and so enabled them to improve their techniques by cutting out traditional stages which the new chemistry showed to be quite unnecessary. Later still, the successful quantification of chemistry that followed John Dalton's atomic theory allowed chemists to calculate the optimum quantities for reactions and so improve manufacturing efficiencies; in the same sort of way mechanical engineers were learning how to optimise the efficiencies of transformation of raw water power and heat power into useful manufacturing power. But it would be

* To some extent the absence of books was also a distinctively English phenomenon. In spite of the importance of mining, no books comparable to Gabriel Jars' *Voyages metallurgiques* (1780) and C. T. Delius' *Anleitung zu der bergbaukunst* (1773) were published in English during the eighteenth century. In the field of mechanical sciences J. T. Desaguliers' magnificent *A Course of Experimental Philosophy* (2 vols, 1734, 1744) went through several editions but subsequent books, such as those by Emerson, Banks and Gregory, showed a fall in standards. A comparison with B. F. de Belidor's *Architecture hydraulique* and R. de Prony's *Nouvelle architecture hydraulique* is very instructive.

wrong to suppose that the impact of the chemical revolution was immediate: for example, W. Richardson's *Chemical Principles of the Metallic Arts*, published in 1790, was expressly designed for the use of manufacturers but was unfortunately vitiated by the author's acceptance of the phlogiston theory. When the book was republished in 1806 it was uncorrected although a reference was made to the 'French' (Lavoisier) system of chemistry.

The pioneers of the new textile industries, Arkwright and his successors, were almost all empirical inventors who had little interest in science and to the extent that they were concerned about education their motives were more often philanthropic than technological. Nevertheless there was, towards the end of the eighteenth century, a wide diffusion of interest in science and innovation among the populace of the midlands and the north where the nascent industries were to be found [*12*]. This was reflected in the correspondence of manufacturers and by the activities of a number of peripatetic science lecturers. The science in question was comparatively elementary and there was no indication of the practice of applied science as we have defined it.

It is hardly surprising, therefore, that there were warning and prophetic voices. In 1782 Thomas Henry foresaw the need for a closer union between science and technics when he complained to the Manchester Literary and Philosophical Society that 'The misfortune is that few dyers are chemists and few chemists are dyers' and, a little later, Thomas Barnes, a Unitarian divine and of the same Society, agreed with Henry and added that '. . . few of our mechanics understand the principles of their own arts and the discoveries made in other collateral and kindred manufactures' [*13*]. But William Jackson, as befitted an ardent believer in progress, was optimistic. In his opinion: '. . . we are doing the drudgery by which the Golden Age is to profit', and, 'Perhaps, some other power may be discovered, as forcible and as manageable as the evaporation from boiling water—another gunpowder that may supersede the present—and other applications of the mechanical powers which may make our present wonders sink into vulgar performances' [*14*]. We would hesitate before describing our own time as a Golden Age, but, considering that Jackson wrote when steam power was in its very infancy, it must be agreed that his prediction was as remarkable as it was correct.

Not, perhaps, unrelated to this mingled optimism for the future and discontent with the present, there occurred, towards the end of that century and in the opening years of the next, the foundation of a number of special societies devoted to the pursuit of science. Partly this movement was due to dissatisfaction with the lethargic Royal Society and with the tyrannical rule of its president, Sir Joseph Banks. Partly it must also have been due to the wider diffusion of scientific knowledge and to an increasing pace of scientific advance. Among the new societies were the Linnean, the Zoological, the Geological, the Astronomical and the Royal Institution; besides these, a number of philosophical societies, like the one in Manchester, also sprang up. The Royal Institution (1799) could, in theory at least, have developed into a technical college as its founder, von Rumford, had intended. This it did not do, although far-seeing people were beginning to predict that such colleges would soon be necessary. At Manchester a group that included Thomas Barnes proposed the foundation of a science and commercial college for the city; the syllabus to include mathematics, belles-lettres, and natural philosophy as well as law, history, commerce and ethics. First place, however, was to be given to chemistry and mechanics because of their importance for manufactures [15]. Barnes had a remarkable appreciation of the possibilities of applied science and he expressly hoped for the discovery of the '*happy art of connecting together liberal science and commercial industry*': words that reveal great prescience [16]. The institution was, in fact, founded and for a number of years lectures were given; notable among them being Thomas Henry's applied chemistry lectures, which included courses on bleaching, dyeing and calico printing. In 1786 the Manchester New College was founded and this institution brought the young John Dalton to the city in 1793.

However, even taking these hopeful signs into consideration, it does not seem that the social institutions of the eighteenth century were well adapted for the advancement of science, either 'pure' or applied. With few opportunities for the practice of applied science and then only on a piecemeal basis, with educational foundations that were ineffective* and with a Royal Society that was little more than a gentleman's club (the social fellows outnumbered the scientific) [17] it is, at first sight, remarkable that the natural sciences could have advanced at all. As if that were not enough, there was also much quackery and unbridled speculation; as, for example, the book published in 1798: 'The sublime Science of Heliography; or the Sun no other than a Body of Ice;

---

* Indeed the development of Gresham's College, which could have led to an early University of London, was actually aborted in the eighteenth century.

overturning all the Received Systems of the Universe'. Yet in compensation for all these drawbacks, the eighteenth-century scientist had the heritage of the 'scientific revolution' and also enjoyed great freedom in his work. This freedom was a great advantage for we still have much to learn of the imaginative process of invention and discovery. The advance of knowledge is not only a function of the intellectual heritage; neither is it automatic or inevitable. Such facts alone justify the very highest degree of intellectual freedom. That constraint, or standardisation, is fatal to science follows from T. H. Huxley's [18] thesis that the advance of natural knowledge has been largely effected by men of opposite, conflicting mind: the one being imaginative and synthetic, aiming at broad, coherent conceptions of the relations between phenomena; the other being positive, critical and analytic.

More specifically we cannot foretell what natural material a creative scientist—for all science is creative—will succeed in co-ordinating within his theories. If the progress of science is towards unification, as I believe the history of its development shows it to be, the scientist will probably have to collect his material from diverse fields of experience. Generally speaking, the greater the theory the wider the fields it will be found to cover and the more diverse the phenomena it will succeed in correlating. Also, we cannot predict the intellectual routes which the scientist may follow, nor the heuristic devices that he may use; Dalton for example, was led to the atomic theory through the study of meteorology and the physics of gases. For this reason it is not surprising that contributions to one branch of science are often made by men whose training has been in other branches. The conclusion we are forced to is that we cannot tell from which quarter enlightenment may come, and it seems very reasonable to suppose that this is still the case to-day.

That scientific freedom was effectively used in the eighteenth century is demonstrated by the substantial advances made in natural knowledge: Herschel discovered the planet Uranus and gauged the shape of the island universe; variable stars were discovered and it was shown that Newtonian laws applied outside the solar system; Bradley discovered the aberration of light and explained its significance; a series of researches carried the new science of electricity up to the verification of the inverse square law, the discovery of the principle of the condenser and of the nature of lighting as well as almost up to the discovery of current electricity and the formulation of circuit constants; Joseph Black took the first steps towards the establishment of a quantified science of heat [*19*], and in chemistry the researches of Black, Priestley and Cavendish made possible the triumphs of Lavoisier and his school.

CONTINENTAL DEVELOPMENTS

It is reasonable to suppose that the distribution of scientific and technical skill is more or less uniform among European peoples. From this it follows that differences in development of 'pure' and applied science must be ascribed to differing social organisations in the various countries. While the science itself is international (see p. 2 above) the way in which—and the extent to which—it is pursued in any particular society must reflect the social patterns of that society. Although generalisations in social and historical matters are apt to be very misleading, we can, perhaps, observe that the bewildering diversity of men engaged in science in England up to recent times was but a reflection of the marked individualism of English life; individualism that has revealed itself in the diversity of religious beliefs and practices as well as in the freedom of English political institutions. We are not surprised, therefore, then we find that English scientists have subscribed to all religions, or to none, that they have come from all walks of life— from pitmen to peers—that their political allegiances have ranged from high Tory to radical revolutionary and that their vocations—when amateurs—have been almost as many as their numbers. It is difficult to believe that any country in the world can rival England in the rich social diversity of its men of science.

(a) France

There had existed in France prior to the events of 1794 a school of artillery at Châlons sur Marne, the Ecole des Ponts et Chaussées (1747), a school of Naval Engineers and, nominally, a School of Mines. The universities of France had long been moribund, and were finally swept away by the Revolution. Despite the inefficient universities, there had been no lack of scientific genius in France; quite the contrary in fact, for the value of French contributions to every branch of science in the eighteenth century is beyond dispute.

Towards the end of the eighteenth century the need for systematic technical and scientific education began to be felt in France as in England. In 1791 Gaspard Monge suggested the establishment of a science school and, at about the same time, Condorcet and Tallyrand put forward a system of education for general engineers. In 1793

Lavoisier laid before the National Convention a comprehensive and liberal scheme for universal education in the arts and sciences [20]. In the following year, J. E. Lamblardie, Director of the Ecole des Ponts et Chausées, obtained, with the enthusiastic support of Monge, a favourable decision from the Committee of Public Safety for the establishment of a school for scientific engineers [21]. The course of instruction was to last three years, students were to be paid by the State and selection was to depend on examination results only. The syllabus included mathematics, 'physics', civil engineering and chemistry together with practical chemistry. The last was adapted from the practice of the mining college which Maria Theresa had established at Schemnitz (now Bǎnska Štiavnica, in Slovakia) in 1770 and which was probably the first technical college in the world.*

In December 1794, six months after the judicial murder of Lavoisier, the new school was opened. It had enrolled 349 students and included among the teaching staff Berthollet, Chaptal, Fourcroy, Monge and Vauquelin. But despite this hopeful beginning the years that followed were difficult for the Ecole Polytechnique. Money was scarce and student numbers fell to 300 and, in 1797, to 200. Political interference followed from the suspicion that the discipline of instruction violated the principle of 'equality'. Conscription was another difficulty: equality demanded that some 90 students should serve as infantry privates. Gradually however the tide began to turn; talent began to show itself among the students. Napoleon supported and protected the school— 39 students accompanied him on the expedition to Egypt—and it became clear that the experiment was justifying itself, for it filled the French armies and public services with efficient engineers. This was hardly surprising for among its professors and students were numbered practically all the leading French savants of the early nineteenth century: Lagrange, Poinsot, Gay-Lussac, Fourier, Sadi Carnot, Fresnel, Ampère, Fourcroy, Malus, Arago, Dulong, Petit, Coriolis and Poisson among others.

The uniqueness of the Ecole Polytechnique lies in the fact that it was the first attempt, anywhere in the world, at a college of applied science.† Its scientific nature is fully guaranteed by the above names;

* The Schemnitz Mining Academy was originally founded at Joachimstal in 1733. It was moved to Prague in 1763 before being finally established at Schemnitz in 1770.

† Discussing Carnot's theoretical analysis of the steam engine, H. T. Pledge remarks [22] that 'it was typical of the difference between the two

its technological bent is clearly indicated by the course of studies. Indeed, it is fair to say that the first major step towards the invention of applied science as a recognised and systematic practice is chiefly attributable to France. The second major step was taken by Germany.

A strong impetus to applied research was provided by the stresses and shortages resulting from war and blockade. The State patronised the Leblanc soda process and encouraged chemists to develop the sugar-beet industries; researches were carried out on gunpowder; new sources of saltpetre were found and new methods of refining it evolved; the best available knowledge was applied to steel making for munitions; chlorine bleaching was popularised and a coffee substitute discovered [23]. J. M. Jacquard invented his loom that could weave complicated repetitive patterns by using punched-card technique: the same process that was to be an essential component of the first mechanical computers. Even in the activity that the English had always prided themselves on, the French were definitely ahead in one important respect: of shipbuilding, G. S. Laird Clowes tells us that up to Trafalgar 'much was learned from captured prizes, for our neighbours had long treated naval architecture from a more scientific standpoint' [24]. Nor should we forget French leadership in such matters as lighter-than-air flight and food preserving (canning). Significant, too, was the introduction of rational units of measurement: the metric system.

Other French achievements at this time were the foundation of the Conservatoire des Arts et Métiers which originated in Vaucanson's collection of machinery (1775) and in the work of a commission of 1793, and the establishment of the Ecole Centrale des Arts et Manufactures (1829), a private enterprise college for the training of factory owners and managers in the principles of scientific industry.

The Revolution seems to have stimulated the talents of a most gifted people [25]. It could, in fact, be reasonably argued that over the fifty years 1775–1825 France produced a galaxy of scientific genius that, *ceteris paribus*, outshines all others from the very beginnings of science up to the present day. The French brought a new touch to science, that of the assured professional. In a way curiously reminiscent of the dashing tactics of the young Napoleonic generals, the French scientists showed themselves able rapidly and efficiently to exploit the possi-

---

nations' that France, rather than England, should first produce the major theory. I suggest that it indicates French superiority in applied science and the great importance of the Ecole Polytechnique in this respect.

bilities in any scientific situation and to follow up initial successes with massive and solid advances. They learned how to work in small teams and the joint paper—'Dulong and Petit', 'Clément and Desormes'— soon became very common. In short, they pioneered the institutions and practices of modern science and went far to transform what had been an affair of amateurs and devotees into a career for highly trained and professional experts. Indeed, if scientific abilities and professional skills coupled with a firm national policy of applying science to industry could have done the trick then the industrial revolution should have taken place in France rather than in England; or at the very least France should soon have overtaken England.

French scientists were assured of well-paid posts in administration, education and even government and there were generous prizes for meritorious research work [26]. But, as Dr. Ben-David has recently pointed out, research was never made an integral and important part of higher education [27], and the great national schools became places of didactic teaching rather than centres of progressive science. There followed, too, an unfortunate development when Napoleon, 'with the power of a despot and the instincts of a drill sergeant'* united the higher educational system of France into one, centralised 'University' (1808). It was geometrical, Cartesian and, in the long run, almost certainly bad for French science. Dr. Ben-David has shown that, for a long time, the effect of centralised administration was to eliminate 'wasteful' duplication of effort and to ensure that for each branch of science and technology there should be only one appropriate institution [28]. But, as we have remarked, experience shows that science flourishes best when there are several centres of excellence in (more or less) friendly competition. What seemed to be sound administrative practice and the best interests of science did not, therefore, coincide. In later years the system was strongly criticised by scientists of the stature of Pasteur and J. B. Dumas. The war of 1870 destroyed the University but some rigidity may have persisted for as late as 1897 the German chemist Ostwald considered that French scientific conservatism was due largely to the educational system.

(b) The German States

Possibly as a consequence of the disastrous Thirty Years War, Germany was, throughout the eighteenth century, a political and intellectual backwater. The 'scientific revolution' and the mechanical philosophy

* As Lyon Playfair later put it.

did not take firm root there and German science remained bogged down in the more mystical notions of the Middle Ages and Renaissance: a tendency that, in some branches of learning, persisted until well into the nineteenth century. Chemistry and metallurgy, to which Germans had made such notable contributions, were forwarded during the eighteenth century by Swedes and Dutchmen, Frenchmen, Scotsmen and Englishmen. Although mining academies were established at Freiberg (1765), Berlin (1770) and Clausthal (1775) besides the one at Schemnitz, the German universities were ineffectual and young Germans with scientific ambitions often sought their fortunes abroad; as, for example, did (Sir William) Herschel and Alexander von Humboldt.

It was at the very nadir of national fortunes, a time of defeat and humiliation following the battle of Jena, that Wilhelm von Humboldt, Alexander's brother, was appointed to direct the Prussian Department of Education. From the day of this appointment we may conveniently date the beginning of the progressive reorganisation of German education. Apart from compulsory and universal primary education there was evolved a comprehensive system of public secondary schools: *Gymnasia, Realschulen* and various trade schools. *Gymnasia* originally restricted their curricula to the classics and mathematics and were the main feeders of the universities. The other schools were 'modern' and technical, teaching science and modern languages but not Greek; they became the main feeders of the technical high schools.*
The pattern of secondary education became, with some local variations, general throughout the German states.

One of the first fruits of von Humboldt's administration was the founding of Berlin University (1809), strangely enough at much the same time that Napoleon was forcing French higher education into its bureaucratic strait-jacket. This turned out to be paradoxical, for although the German universities were State institutions, maintained out of public funds, each with its State representative—the Curator— and although the professors were recognised officials of authoritarian governments, they yet retained remarkable freedom. It was their boast, and it seems to have been justified, that in Germany every scholar was a professor and every professor a scholar. They believed, and held to their belief, that two freedoms were essential for the proper functioning

* It is worth noting that many German scientists of the first rank later strongly recommended a classical education as the best preliminary training for the would-be scientist.

of a university: freedom of teaching and freedom of learning (*Lehr-und Lernfreiheit*). No doubt there were occasional stresses and strains and sometimes there was State interference, as in the case of the seven Professors of Göttingen (1837); but, these things apart, the universities enjoyed such steady progress throughout the nineteenth century that, by the beginning of the twentieth they could fairly claim to be the best in Europe. An accidental advantage enjoyed by the German universities was that, thanks to the politically fragmented state of the country prior to 1871, there was never any danger of overcentralisation. Each German state was responsible for its own university—or universities—and a number of competing centres of excellence could thus develop unhindered during the nineteenth century.

The German universities were professional schools rather than places of liberal education; and it was as such that they were fostered by a relatively enlightened bureaucracy [*29*]. Most of the students were preparing for State examinations for entry to the various civil services, the churches, the law, medicine and teaching, and by no means all of them were degree candidates. With one or two minor exceptions the only degree awarded was the doctorate of philosophy (Ph.D.). The usual practice of those who aspired to this honour was to attend university lectures for the first year or so and then, in the remaining years, to undertake some original research under the professor's direction. On the completion of this work a thesis was submitted and the candidate had to undergo a gruelling oral examination of some hours' duration. A consequence of this practice was that discipline, so strict in the schools, was very relaxed at the universities. In accordance with *Lernfreiheit* attendance at lectures was not compulsory and the student was not harassed by written examinations. It will be observed that the German system whereby the award of a degree was conditional upon original work and oral examination was much closer in spirit to the ideals of the medieval university than was the contemporaneous English custom of awarding degrees to those who passed written examinations.

Apart from charges of infidelity and 'rationalism' the main criticism of German universities made by Englishmen was that they tended to produce narrow specialists. Thus an English reviewer wrote, in 1839: 'The Germans are become the most learned men in the world, but the least manly, the least capable of being members of free and independent communities. . . . Moreover, as a general rule to which there are few exceptions, beyond the limits of his own particular study the German professor is profoundly, nay ludicrously ignorant' [30]. To take a more moderate example; an early admirer of the German universities, Walter Perry, writing in 1846, agreed that we tended to think of the German professor as devoted to special studies: one who wrote tremendous theses on small subjects [31]. But it is probably nearer the truth to credit the German universities with awareness of the dangers of over-specialisation and with reasoned attempts, within the framework of *Lernfreiheit* and external demand, to discourage it. Examples of this attitude were not lacking: in the mid-century period A. W. Hofmann stated explicitly that German universities disapproved of 'one-sidedness'; later on, Ostwald pointed out that chemistry students were encouraged to attend history and philosophy lectures in their first year at university [32], while just before the 1914 war the educationist Friedrich Paulsen eloquently restated the case against excessive specialisation and suggested ways of combating it.

At the beginning of the nineteenth century the position of experi-mental science in Germany was not satisfactory. The anti-scientific philosophy of Hegel was then dominant and the quasi-mystical 'science' of Lorenz Oken and Wilbrand tended to denigrate experi-mental work. In fact the first generation of German chemists, men like Mitscherlich, Gustav Rose, Wöhler and Liebig were constrained to go abroad for their education—the first three went to Sweden to sit at the feet of Berzelius while Liebig went to Paris and studied under Gay-Lussac [33].* It is to Liebig, more than to anyone else, that the subse-quent superiority of German university science is due, for from 1825 onwards he developed at the minor University of Giessen the famous laboratory that was to be the *fons et origo* of university research labora-tories and, only slightly less directly, of the great industrial research laboratories in Germany and elsewhere at the end of the century. The physics laboratory was a much later development and was one that England, as much as any other country, can claim to have pioneered.

To sum up briefly, it is impossible to deny that the ideal motivating the German universities was one of disinterested love of learning, not only in science but in other fields of knowledge as well; fields whose practical utility was negligible. A love of learning for its own sake was not peculiar to the Germans, far from it, but they seem to have suc-ceeded in organising it, without sacrificing the essential freedoms, more effectively than any of their contemporaries.

At the end of the Napoleonic wars the demand for technical educa-

* No German scientists seem to have come to England to study.

tion was raised in Germany as elsewhere [*34*]. It arose partly from national requirements for technicians in the transport, forestry and metallurgical industries, and partly from envy of the industrial successes of Great Britain. For long the latter country had been isolated from Europe and when, to quote a Royal Commission of 1884, '. . . the continental countries began to construct railways, erect modern mills and mechanical workshops, they found themselves face to face with a full-grown industrial organisation in this country which was almost a sealed book to those who could not obtain access to our factories'. Inspired to a considerable extent by the Ecole Polytechnique, technical schools were founded in Germany with the intention of injecting science into the workshops. The Berlin Gewerbeschule dates from 1821 and similar schools were opened in Darmstadt (1822), Karlsruhe (1825), München (1827), Dresden (1828), Nürnberg (1829), Kassel (1830), Hannover (1831) and Augsburg (1833). In fact, before many years had passed every large town in Germany had its polytechnic school. At first intended for youths of 15 or so, and without consideration as to whether they were to be managers or foremen, these colleges were progressively raised, by attentive governments, towards university status. The standard required for admission was steadily improved, greater self-government was achieved, *Lernfreiheit* was introduced and ultimate control was vested in the Kultusministerium. Between 1860 and 1890 these colleges, with the addition of ones established in towns like Aachen, Braunschweig and Stuttgart, were raised to Technical High School status. In 1879 the Berlin Gewerbeschule was amalgamated with the Bauakademie (1799); and finally in 1899, the centenary of the latter, the Kaiser raised them all to the status of degree-awarding universities. The older universities did not, we infer, look upon any of these developments with warm approval. But by the end of the nineteenth century the Technische Hochschulen were of immense importance to the German economy and were in process of rapid expansion. They were, in accordance with German practice, State institutions from the very first. In this respect they, and the older universities, stood in marked contrast to English practice; while the freedom they enjoyed distinguished them sharply from French institutions.

A COMPARISON

The German universities were, to a far greater extent than those of England, truly national institutions. From the beginning of the nineteenth century onwards, during the years when Germany was struggling to attain unity, the universities had, for all their early shortcomings, embodied national aspirations and conserved national culture in a way that no other institutions could have done. England, on the other hand, sure of her national identity and confirmed in her political unity, could afford to neglect the two universities; hence the decline in standards during the eighteenth century and the slow rate of progress in the nineteenth.

A better comparison is between German and Scottish universities. The Scots were, with good reason, conscious of the need to preserve national identity and to conserve national culture; moreover they traditionally reverenced learning for its own sake. Accordingly their universities were, by common consent, the best in Europe during the eighteenth century [35]; significantly, this was after the Act of Union with England. Glasgow and Edinburgh universities were particularly famous for chemistry teaching, due in great measure to the abilities of Joseph Black and, in the case of Edinburgh, his successor there, T. C. Hope. Audiences of up to 500 or more were quite common at chemistry lectures and it is reasonable to wonder why, with such a strong background, scientific education in Scotland did not evolve schools of scientific research that might, in turn, have led to organised applied science. However, J. B. Morrell has demonstrated that, in the case of Edinburgh University, chemistry was taught solely through the medium of lectures and attempts to establish practical classes failed. For this the blame lay with the administrative arrangements of Scottish universities and in particular with the custom of remunerating the professor solely from the fees paid by those attending his lectures. The consequence was that, in the absence of practical classes, scientific research could not become part of university education in Scotland [36].

REFERENCES

[1] D. S. L. Cardwell, 'Early Development of the Concepts of Power, Work and Energy', *British Journal for the History of Science* (June 1967) iii, p. 209.

[2] D. S. L. Cardwell, *Steam Power in the Eighteenth Century* (Sheed and Ward, London 1963) and *From Watt to Clausius: the Rise of Thermodynamics in the Early Industrial Age* (Heinemann, London 1971).

[3] *Dictionary of National Biography.*

[4] W. Walker, *Memoirs of the Distinguished Men of Science of Great Britain Living in 1807–8* (London 1862).

[5] Ludwig Darmstaedter, *Handbuch zur geschichte der naturwissenschaften und der technik* (Berlin 1907).

[6] A. D. Godley, *Oxford in the Eighteenth Century* (London 1908).

[7] Adam Smith, *The Wealth of Nations* (6th edition, Methuen, London 1950) ii, p. 249.

[8] William Whewell, Royal Institution Lecture (London 1855).

[9] D. A. Winstanley, *Unreformed Cambridge* (Cambridge University Press, 1935), pp. 318–30.

[10] H. McLachlan, *English Education under the Test Acts* (University of Manchester Historical Series, No. 59, 1931) appendices.

[11] Hermann Boerhaave's *Elementa chemiae* first appeared in English in a pirate edition in 1727: *A New Method of Chemistry*, translated by Peter Shaw. The authorised edition of 1735 was translated by Thomas Dallowe. A further Shaw edition was published in 1741.

[12] A. E. Musson and E. Robinson, 'Science and Industry in the Late Eighteenth Century', *Economic History Review* (1960) xiii, p. 238.

[13] Thomas Barnes, 'On the Affinity Subsisting Between the Arts', *Memoirs of the Manchester Literary and Philosophical Society* (1785) i, pp. 72–89.

[14] William Jackson, *The Four Ages* (London 1798) pp. 60, 92. See also Sir John Herschel, *Preliminary Discourse on Natural Philosophy* (London 1831) p. 359.

[15] Thomas Barnes, 'Proposals for Establishing in Manchester, a Plan of Liberal Education . . .', *Manchester Memoirs* (1785) ii, pp. 30–41.

[16] Thomas Barnes, 'A Plan for the Improvement and Extension of Liberal Education in Manchester', *Manchester Memoirs* (1785) ii, p. 16.

[17] See D. M. Stimson, *Scientists and Amateurs* (Schumann, New York 1948) and Sir Henry Lyons, *The Royal Society, 1660–1940* (Cambridge University Press, 1944).

[18] T. H. Huxley, Prefatory note to the English edition of Ernst Haeckel's *Freedom in Science and Teaching* (London 1879).

[19] For Black's contribution to the quantification of theories of heat see D. S. L. Cardwell, op. cit. [2].

[20] D. McKie, *Antoine Lavoisier, the Father of Chemistry* (Victor Gollancz, London 1935).

[21] For practical chemistry teaching in the early days of the Ecole Polytechnique see G. Pinet, *Histoire de l'Ecole Polytechnique* (Paris 1887) p. 366; also the papers by W. A. Smeaton in

Annals of Science (1954) x, p. 224; (1955) xi, pp. 257, 309; (1965) xxi, p. 33 and F. B. Artz, *The Development of Technical Education in France 1500–1850* (M.I.T. Press, Cambridge, Mass. 1966).

[22] H. T. Pledge, *Science since 1500* (H.M.S.O., London 1939) pp. 143–4.

[23] J. H. Clapham, *The Economic Development of France and Germany, 1815–1914* (Cambridge University Press, 1936).

[24] G. S. Laird Clowes, *Sailing Ships: Their History and Development* (H.M.S.O., London 1932) p. 93.

[25] The achievements of French science during the revolutionary period have recently received considerable attention from historians. Besides the works of McKie and Smeaton, referred to above, see M. P. Crosland, *The Society of Arcueil* (Heinemann, London 1967) and Robert Fox, *The Caloric Theory of Gases from Montgolfier to Regnault* (Oxford University Press, 1971).

[26] M. P. Crosland, op. cit., [25].

[27] Joseph Ben-David, 'The Rise and Decline of France as a Scientific Centre', *Minerva* (April 1970) viii, p. 160.

[28] Ibid.

[29] Joseph Ben-David and Abraham Zloczower, 'Universities and Academic Systems in Modern Societies', *European Journal of Sociology* (1962) iii, p. 45.

[30] Anon, 'University Reform', *British and Foreign Review* (1837) ix.

[31] Walter Perry, *German University Education* (London 1846).

[32] Quoted by J. Norman Lockyer, *Education and National Progress: essays and addresses, 1870–1905* (London 1906).

[33] W. A. Shenstone, *Justus von Liebig* (London 1895).

[34] A. E. Twentyman, 'A Note on the Early History of Technical High Schools in Germany', *Special Reports on Educational, Subjects,* Cd. 836 (London 1902) ix.

[35] S. D'Irsay, *Histoire des Universités* (Paris 1935) ii, p. 290 et seq.

[36] J. B. Morrell, 'Practical Chemistry in the University of Edinburgh 1799–1843', *Ambix* (July 1969) xvi, p. 66.

CHAPTER THREE

English Developments:
1800–40

At the beginning of the nineteenth century England was engaged in a desperate war with France; she was also irrevocably committed to that memorable enterprise, the industrial revolution. The evolutionary development of Watt's steam engine and the skills of English mechanics made possible the transfer of the pace-making textile mills from country districts—where abundant water power was available—to towns* where labour, markets and service and collateral industries were to be found [1]. Contemporaneously, as Samuel Smiles indicated, the great executive engineers, such as Telford, Rennie and Stephenson, were revolutionising transport by road, canal and rail so that a country which had been in many places an impassable wilderness was rapidly tamed and made suitable for industrialisation [2]. These dramatic events have tended, such is the common historical perspective, to overshadow the equally important if less immediately striking series of changes that constituted, in effect, a revolution in English agriculture. It is self-evident, that without a great leap forward in agriculture, it would have been impossible to feed the rapidly growing populations of the new industrial towns as well as, at the same time, victualling the great armies and fleets that Britain had to maintain. Agriculture is, to-day, Britain's most efficient industry; there are grounds for believing that this has also been the case several times in the past.

Professor Armytage has pointed out that Englishmen have long been deeply concerned with the world of organic nature. This concern is expressed, for example, in the acknowledged excellence of the English garden and is even reflected in the imagery of English language [3]. What were later called colonies were originally known as 'plantations' and with them were associated, as Professor Armytage puts it, a 'plantocracy', corresponding to the 'technocracy' of modern times. Kew Gardens were, at the beginning of the nineteenth century, un-

* R. L. Hills points out that abundant water was still required by steam-driven mills even when they were located in towns. What was no longer necessary was a considerable head, or fall of water to produce power.

rivalled in Europe and the pursuit of botany enjoyed official favour and support, albeit that it was hardly a science in those days before Darwin and Mendel. But botany was only an intellectual extension, as it were, of the great business of agriculture and in this field the eighteenth century had seen a succession of notable improvers and innovators—Tull, Townshend, Coke, Bakewell and Arthur Young. The first four of these men had introduced new machines, new methods of marling the soil, new crops, and new principles of stock breeding while the fifth, Arthur Young who was made Secretary of the Board of Agriculture when it was first established in 1793, was responsible for bringing more land into much more effective use [4]. Agriculture, then, was efficient and favoured by those in authority; almost inevitably, it seems, it acquired a scientific component: in 1803 the popular and able Humphry Davy published his lectures on agricultural chemistry. These inaugurated a new branch of science that was later to arouse the interest of Liebig and play an important, if indirect, part in the development of organised science in England.

We can, in fact, consider the development of science in England during the nineteenth century as taking place in three distinct phases. The first began after Waterloo and was notable for the widespread interest in science shown by the artisan class and the consequent founding of the mechanics' institutes. The second phase occurred in mid-century and is associated with the general spread of the examination system, the founding of the Science and Art Department, the Exhibition of 1851 and the personal efforts of such men as the Prince Consort and Lyon Playfair. The third phase began in the 1880s and was not completed by the time the Great War broke out; to attempt to sum it up in a few words would be impossible. Between these phases there were periods of relative stagnation; as if each wave of enterprise had exhausted itself.

The opening years of the new century were marked by important scientific developments. On the one hand there were the advances associated with the names of Dalton, Thomas Young, Davy, Wollaston and others, while on the other hand, there was the proliferation of societies and philosophical institutes; activities which indicated a considerable interest in science on the part of the middle classes. Yet this interest was, we may infer, largely amateur, even dilettante, for, over the period, there was only one public research laboratory, and that was at the Royal Institution.

The popularisation of science was greatly aided by the public status and influence of such men as Davy, Rumford, Wollaston, and Young, and confirmed by the very evident effects of mechanical revolution. But the amateur tradition continued until late into the new century; indeed, it was not until well into the twentieth century that the last amateur died. Men like Babbington, Snow Harris, Francis Ronalds (one of the first inventors of the electric telegraph), the Childrens, Luke Howard, Francis Baily, Edward Howard, and, later, Warren de la Rue, W. R. Grove, Q.C., Lord Wrottesley, the Earl of Rosse, General Sabine, Neill Arnott, J. P. Gassiot, William Spottiswoode, and many others were amateurs. Even some of the very great—Dalton, Darwin, James Prescott Joule—were of the class of devotees, and Faraday himself can be considered one: did he not turn down a lucrative career as a consulting chemist in order to continue with his independent researches?

At the beginning of the period the diffusion of scientific knowledge was to take place within a highly differentiated class structure complicated by a variety of religious differences. At its best the scheme of social rights and duties of the upper classes had decreed that, for the poor, '. . . their morality and religion should be provided for them by their superiors, who should see them properly taught it, and should do all that is necessary to ensure their being, in return for their labour and attachment, properly fed, clothed, housed, spiritually edified and innocently amused'.* But this period was one of war, or repression and revolt, and it was widely believed by the fortunate that to educate the working classes would be to invite disaster. In accordance with the prevailing social philosophy, the well-known scientist, Davies Gilbert later P.R.S., was among those who had, as an M.P., bitterly and uncompromisingly opposed Whitbread's Parochial Schools Bill of 1807 [5].† Patrick Colquhoun, the magistrate and police reformer who had been associated with Quaker relief work, would allow an elementary education with due emphasis on morals and proper subordination, but 'science and learning, if universally diffused, would speedily overturn the best constituted government on earth' [6]. The fear of science is interesting; it sprang directly from the course of events in France.

The real diffusion of science began first to be effective in the more peaceful 1820s. It was forwarded by the Benthamites and by the very

* John Stuart Mill.
† But Giddy deserves great credit for his patronage of Jonathan Hornblower, Richard Trevithick and Humphry Davy.

progressive school in Edinburgh that had, as its mouthpiece, the *Edinburgh Review*. In the Scottish capital in the closing years of the eighteenth century there was a characteristically brilliant constellation of talent—John Playfair, Sydney Smith, Walter Scott, Dugald Stewart, Francis Jeffrey, Henry Brougham, George Birkbeck and many others. The scientific element was strong in this school, even among those whose main interests lay elsewhere: thus the lawyer Brougham had had the good fortune to have attended Joseph Black's lectures. Towards the end of his long life Brougham remembered [7] that Black had given him greater intellectual gratification than had Pitt, Fox, or any other orator or lecturer he had heard. Another interesting member of the Edinburgh school was one of the first lady-scientists: Mary Somerville. An able mathematician, she translated Laplace's *Mécanique celeste* for Brougham and wrote several other books of a scientific nature [8].

This, however, is by the way. Roughly correlated with class levels there were three distinct responses to the scientific and educational movement for the upper classes, the first glimmerings of effective university reform: for the middle classes, the foundation of London and Durham universities; for the working classes there was that remarkable movement—the rise of mechanics' institutes.

The first objective study of the mechanisation of industry—the core of the industrial revolution—was a remarkable book by Charles Babbage, *The Economy of Machinery and Manufactures*, published in 1832. This was followed by Andrew Ure's *Philosophy of Manufactures* (London 1835) which was in some respects a fairly obvious imitation of Babbage's work and which was marred by ludicrous eulogies (to the modern reader, at least) of the delights of labour in the new cotton mills coupled with indiscriminate attacks on the critics of the factory system. Nevertheless Ure did make some penetrating and original observations. He understood the importance of automatic, self-controlling machinery and he had some insight into the possibility of an automatic factory. He pointed out that, with the rise of the specialised machinery makers, in the textile industries, at least, machines commonly became obsolete before they wore out. And he realised that these remarkable developments were only possible because there was a considerable population of highly skilled labour in the northern industrial towns: 'It is this skill in machine-mounting or adjusting . . . which gives to our factories not merely their existing superiority over foreign rivalry but the best security for its permanence. Indeed the concentration of mechanised talent and activity in the districts of

Manchester and Leeds is indescribable by the pen and must be studied confidentially behind the scenes in order to be duly understood or appreciated.' Such communities of specialised labour represented perhaps the working-class élite of the industrial revolution: increasingly affluent, skilled and industrious, these men had had little or no formal education. They were anxious to know about science, technology and industry; about the new world they were creating and the progress they were ensuring. Their means of achieving these ends were to be in the first instance, the mechanics' institutes.

The Birmingham Sunday Society, founded in 1789, has been credited with being the first attempt at a mechanics' institute. In 1796 it changed its name to the Brotherly Society and had, for an object, the diffusion of a taste for science among the Birmingham artisans [9]. George Birkbeck knew of this society and when, after his appointment as professor of natural philosophy at the Andersonian Institution in Glasgow, he commenced his famous lectures for artisans he may well have had the Birmingham society in his mind. The success of Birkbeck's lectures was such that when he left Glasgow in 1804 to take up medical practice in London they were continued by his successor, Andrew Ure. These lectures, and the mechanics' library which had been founded at the Andersonian, may be said to have constituted the first of the mechanics' institutes. But the origination of the institutes can hardly be ascribed to one venture, still less to one man—however much credit is due to Birkbeck and those associated with him. Such a vast movement could hardly be conjured up by one individual agency.

Although a Timothy Claxton had founded a Mechanical Institution in 1817 in London, it did not survive more than two years, and seems to have been quite unconnected with Birkbeck's enterprise. On the other hand the first mechanics' institute to be formally founded was the Edinburgh School of Arts (April 1821). It originated with a group of wealthy and benevolent citizens, notable among whom was Leonard Horner,* F.R.S., and was admittedly inspired by the Glasgow experiment. This School, which later became the Heriot-Watt College, and finally, the Heriot-Watt University, offered courses in chemistry and natural philosophy and in applied chemistry and applied mechanics. The Glasgow mechanics naturally took note of this and, evidently discontented with the subordinate position they occupied in the

* Horner, a member of the Edinburgh group, was later to be the first Warden of University College, London.

Andersonian, finally broke with that Institution and established their own Institute in July 1823. It was not long before these Scottish experiments were observed in England, especially by the radical elements. By November 1823 Thomas Hodgskin and Joseph Robertson of the London *Mechanics' Magazine* were in correspondence with Birkbeck on the subject of an Institute for London. As a result of this Birkbeck was drawn into a new enterprise and the London Mechanics' Institute (now Birkbeck College) was established in December 1823. At first it was comparatively well endowed; it had the approval and support of distinguished radicals and the Royal Duke of Sussex showed some interest in it. The first lecture syllabuses included chemistry, mathematics, hydrostatics, applied chemistry, astronomy and electricity.

After this the movement spread with astonishing speed. Before 1826 every large town and many a small one had its institute, and several of these had memberships of hundreds with a few topping the thousand mark. Institutes were founded in Bermondsey, Spitalfields, Hackney, Rotherhithe, Southwark, Chiswick, and Kensington as well as at Manchester, Liverpool, Sheffield, Leeds, Newcastle (George Stephenson was a V.P.), Carlisle, Hull, Darlington (the Bishop of Durham was President), South Shields, Plymouth, Bristol, Ipswich, etc., and even in such places as Kendal, Hawick and Bath. Sometimes the initiative came from the benevolent among the rich and sometimes it came from the workers themselves, but always it was entirely voluntary and for it neither the State nor the recognised educational institutions of the country could claim the slightest credit.

In the van of the movement were the dissenting clergy and dissenters generally; while strong support came from men like James Mill, Cobbett, Grote, Ricardo, Place, J. C. Hobhouse, Sir Francis Burdett and Henry Brougham. Brougham's support was far from passive; in fact, he flung down the disturbing challenge that the self-nominated superiors of society needed a little competition from below: '. . . we make a pleasure of business, while they make a business of pleasure'. Sentiments like this, hardly calculated to soothe the fears of the fortunate, aroused the opposition forces. The high Tory press was bitterly hostile (e.g. the *Courier*, the *St. James's Chronicle* and *John Bull*), while pamphleteers like 'Country Gentleman' [10] and the Rev. E. W. Grinfield expressed the old fears of science [11]. The conservative *Quarterly Review* remarked, somewhat aloofly, that both the benefits and the dangers had been exaggerated. It was not important to teach science; what should be done was to teach some economics—the iron laws and the immutable causes of the inequality of man [*12*].

Privilege, generally, was very uneasy about the new development; the high aristocracy and the Anglican clergy were opposed to the movement and did all they could to hinder it. One of these opponents, a gentleman and a scholar, had the moral courage to lecture the mechanics on the folly of their institutes. His arguments were stale but he did make one remarkable observation. The Southwark mechanics were to have a chemical laboratory—but, 'in first-rate scientific institutions, attended by men of leisure, rank and education, the laboratories are seldom used'. [*13*]

In retrospect it is evident that there were several components in this movement. There were the educational and social ideals of the Benthamites together with the early impetus to combination and trades-unionism; for, although discussions about politics and religion were banned from most institutes (for reasons that are sufficiently obvious), it is difficult to believe that there was no relationship between the movement and the spread of radical politics: witness the parts played by men like Hodgskin, Robertson and Rowland Detrosier. There was also the need for relaxation and social life in the busy new industrial towns: for the literate artisans the institutes provided libraries and lectures; for all they provided entertainment and the opportunity for social intercourse. But perhaps the most remarkable feature of this remarkable movement was the prime importance attached to the natural sciences. That large numbers of hardworking artisans were prepared to attend scientific lectures, is clear enough proof of the high regard for science; a sure indication of progress which caused the *Edinburgh Review* to enthuse 'The sacred thirst for science is becoming epidemic'. As Delisle Burns remarked, at a time when the universities did next to nothing the institutes ministered to all classes in this need [*14*].

When we come to examine the lectures, syllabuses and aims of the institutes the picture becomes a little obscure. The founders of the institutes seem to have hoped that they would achieve four objectives: firstly, the injection of science into the workshops of the country with consequent economic benefit. Indeed on Baconian grounds an increase in the numbers of the scientifically instructed must lead to an increase in the number of science-based inventions. There was, therefore, the dazzling prospect of an avalanche, a cumulative avalanche, of science-based inventions as scientific knowledge spread among the skilled artisans and mechanics of the country [*15*]. Secondly, the wider diffusion of science and rational knowledge would, it was hoped, banish superstition and ignorance. Thirdly, the movement was expected to hasten the progress of industry by increasing the numbers of those able to pursue it; and, lastly, in accordance with the views of Adam Smith they hoped that science and education would offset the degrading effects of the industrial division of labour.

The leaders of the new movement were, of course, deeply, if not clearly, aware of the important effects that science could have on technology and hence on social conditions—the collaboration between Joseph Black and James Watt was frequently emphasised. 'Is the firm and powerful voice of science to pervade the workshops of the kingdom,' asked Birkbeck, 'or shall the feeble and uncertain vote of experience continue to prevail?' [*16*]. Addressing the Manchester mechanics in 1825 (Sir) Benjamin Heywood emphasised the science that underlay the arts, instancing specifically Watt, Hargreaves* and Murdock [*17*]. Brougham, too, in his *Practical Observations . . . on the Education of the People* noted the intimate connection subsisting between the mechanical philosophy and the arts [*18*]. At a later date Detrosier put the proposition in general terms, although perhaps from a different political viewpoint: knowledge is power. Knowledge has raised man up. Man is progressive, so are social institutions [*19*].

It is apparent from the preambles to the deeds of foundation of the institutes, that to teach the 'sciences underlying the arts' was the prime object. In practice this meant largely the teaching of the 'pure' sciences: mathematics, chemistry, physics, astronomy, botany, meteorology, the theory of the steam engine etc. Trades were not generally taught; it was felt that the proper place to learn them, and so to acquire skills, was the workshop. In pursuit of this policy some of the larger institutes had chemistry laboratories and some were able so to organise their teaching that systematic lecture courses, limited to individual sciences, could be given. In fact most, if not all, the sciences seem to have been taught in the largest institutes. But admirable as all this was, no precise meaning can be given to the expression 'sciences underlying the arts'. A brief mental inspection of the diversity of the arts and the complexity of the sciences will show how vague this expression is. In some specific instances the connection is fairly clear; as for example, in the case of the chemical industry. But generally the relationship is more complex.

* This was an unfortunate example. Hargreaves was an empirical inventor and there was no evidence that he knew anything at all about science.

The sciences 'underlying' the textile industries, for example, are several; and between the 'pure' science, on the one hand, and the practical art on the other, the connection is often far from obvious. The reason for this is not difficult to discover. Even the most stalwart upholder of theories of the economic motivation of science must agree that no one science is derived from one particular art. No one denies the importance of technology in the rise of the sciences, but in so far as a science derives its data and its problems from the industrial arts, it does so from several of the arts—not from one. Even chemistry, perhaps the most 'practical' of all the sciences, and certainly the first one to be widely applied, owes much to activities quite distinct from the chemical industry (medicine, for example) and as the structure of the science develops those portions of it that may be directly relevant to a particular art will become more and more mingled with extraneous elements derived, perhaps, from other arts. To put the point briefly, nature is not organised to suit the differences between the trades and arts; you cannot learn, for example, dyeing chemistry without knowing 'pure' chemistry. There is nothing remarkable about this for science, as has already been pointed out, progresses by achieving greater generality. Perhaps this is what William Whewell meant when he remarked, rather quaintly: 'Art was the mother of science; the vigorous and comely mother of a daughter of far loftier and serener beauty.'

Thus the industrial improvement that was looked for was rather a general rise in scientific knowledge and a spread of scientific habits of thought than the specific application of science to given problems. Technical education, or the teaching of a particular trade together with the relevant sciences, was not generally attempted; still less was applied science contemplated. The latter was fostered neither by government nor by the universities nor by industry. But in the meanwhile the institute movement spread over the world, reaching America and even the shores of the Pacific and Indian Oceans. Baron Charles Dupin, who had visited Birkbeck and had noted the progress made, took the idea to France, where, with State aid, it proved successful.

Eventually, for reasons that will be discussed later, the mechanics' institutes failed; first as *mechanics'* institutes and then completely as institutes. But they had not been without value: as the biographer of Birkbeck remarked: 'When the pent up waters of science were set free a new stream of life traversed the country in all directions. And that stream has never ceased to flow' [20].

MIDDLE-CLASS EDUCATION

A resultant of the forces of Benthamism and the Edinburgh school was the foundation of University College, London.* This institution, of which the moving spirits were Thomas Campbell, Bentham and Brougham, was organised on Scottish and German lines, in contrast with the practice of the English universities, and it has as its purpose the satisfaction of the needs of the educationally deprived middle classes, of those excluded from the universities. Although this was surely innocent enough, the experiment aroused the wrath of the Anglican Tory elements. Not that there was any fear that the middle classes would become seditious: that would have been preposterous. The trouble was that it was found impossible in practice, to reconcile a 'university', open on equal terms to Anglicans, dissenters, Roman Catholics, Jews and free-thinkers, with the study of divinity and the holding of services, and it was this absence of religious instruction that rallied the opposing forces of orthodoxy. The critics held that because a university education had hitherto implied the education of a Christian gentleman, the 'university' was clearly a most dangerous step in the direction of infidelity and atheism.† The debate was conducted with some acerbity on both sides, but the outcome, surprisingly, proved beneficial: whatever merits, or otherwise, there may have been in the arguments of the opposing parties, the Anglicans, finding they could not prevent the foundation of University College, took the rational and sensible step of founding a rival college—King's. Something of this nature had already been suggested by 'Christianus' (the Rev. George D'Oyly) in his letter to Sir Robert Peel protesting against the establishment of the infidel 'university' [21]. D'Oyly had also suggested the founding of a (Christian) college in a northern industrial town and, in the event, this latter suggestion, too, was carried out for, in 1836, the University of Durham was founded. Just as there had been embryonic university foundations in London long before the nineteenth century—Gresham College for example—so also Durham could claim a history of past attempts at higher education; under the Commonwealth a college was founded in that city, but the Restoration had killed the enterprise. The nineteenth-century Durham University

* H. Hale Bellott, *The History of University College, London, 1826–1926* (University of London, 1928).

† This argument was, as Sir William Hamilton pointed out, unsound; religious instruction had never, up to the end of the eighteenth century, formed part of a general university education.

was to prove a hardier plant although its earlier history was not one of remarkable progress. It was modelled very closely on the Oxford and Cambridge of the time, and served as a local university for the wealthier classes at a time when travel was still a matter of some difficulty. But the development of the railways, bringing Oxford and Cambridge much closer, coupled with an adherence to a system of university education rapidly passing out of favour even among the classes for whom it was designed, were factors that were to make great difficulties for Durham in the middle years of the century.

From the first the new 'University of London' included in its syllabus not only the classics and mathematics plus theoretical physics as sanctioned by the universities, but also the progressive sciences of chemistry, experimental philosophy (physics), botany, economics, geography, etc.—subjects that were commonly neglected by the universities. Among the first professors appointed were Henry Malden in classics and Augustus de Morgan in mathematics, both being— appropriately—distinguished Cambridge alumni. De Morgan, who made contributions to mathematical logic, was an extremely clear thinker with a very wide range of interests; his writings on the history of science are conspicuous for both scholarship and wit. The chairs of chemistry and physics were held by Edward Turner and Dr. Dionysus Lardner respectively, while the professor of botany was James Lindley.

Prospective students were advised to take a four-year course of studies, comprising classics, mathematics, physics, chemistry, economics, philosophy and jurisprudence, on the conclusion of which, subject to satisfactory conduct and approved examination, a certificate of proficiency was awarded [22]; this, it was considered, ensured adequately 'liberal' education. But provision was also made for two other types of training: there was professional education with three main branches—law, medicine and engineering and, secondly, a formal education, amusingly termed 'ornamental accomplishments', that embraced Italian, Spanish and German literature together with botany and, oddly enough, geology and zoology [23]. Whatever we may think of this latter mode of education the syllabuses show quite clearly that the college made no attempt to turn out specialists, and learning was, consequently, quite free. It was also notable that, as regards science, the subjects taught were very up-to-date, the latest knowledge being incorporated in the syllabuses; the study of the steam engine was included in the physics course and, in mathematics, analytical methods were used. It was far-sighted, too, to attempt to teach a form of 'applied science': from the day of its foundation the college had a professor of engineering; the first being that John Millington who had shown himself such a good friend of the mechanics' institute movement. His colleague, Lardner, took a hand in this work for he gave instruction in geodesy and surveying. Even this does not exhaust the pioneer work undertaken, for, from the first, the college provided instruction in practical chemistry.

In one sense University College was residuary legatee of the dissenting academies of the previous century. But the basis of its support was much broader than those establishments had enjoyed. From the lists of subscribers to the two new colleges we can infer the polarisation of two social forces: Anglican-Tory and radical-dissent. To University College contributed the cream of the Philosophical Radicals: Bentham, Mill, Brougham, Grote, etc., and many leading members of the dissenting, Roman Catholic and Jewish communities. The King's College supporters, on the other hand, were led by the Duke of Wellington and a phalanx of some thirty-two archbishops and bishops of the Establishment, strongly supported by the Heads and Fellows of Oxford and Cambridge colleges. A measure of the ecclesiastical influence in the college was that members of staff were required to be members of the Anglican communion; although the same was not demanded of students.*

The syllabuses show that the courses of instruction at King's College were fully as liberal as those given at University College with, of course, the addition of religious instruction. The purpose of the college was announced as two-fold: to prepare for commercial and professional life and to give preliminary training to those who wished to go on to the universities to read for a degree. To carry out this programme King's like University College, recruited an excellent staff. One of the first to be appointed was the Rev. Henry Moseley, Professor of Natural Philosophy who had been seventh Wrangler at Cambridge.† J. F. Daniell was elected professor of chemistry, while geology boasted a professor who was perhaps the most distinguished of all—Charles Lyell. A few years later Sir Charles Wheatstone was elected Professor of Experimental Philosophy.

* F. J. C. Hearnshaw, *Centenary History of King's College, London, 1828-1928* (Harrap, London 1929).

† Moseley later played a notable part in the technical education movement. See below, page 116.

There was a slight difference between the two colleges which reflected the differing origins. The students at King's were generally younger than those at University College and frequently went on, as intended, to complete their education at Oxford or Cambridge. King's was, in effect, at that time more of a high school than a university college. But a more significant difference was the prevailing ecclesiasticism of King's which at staff level, was soon to make itself felt to the disadvantage of the college. For when Lyell published the highly controversial second volume of his *Principles of Geology* (1832) it constituted, or was thought to constitute, a challenge to religious orthodoxy which made his position on the staff paradoxical to say the least. Consequently, the college soon lost its highly distinguished—to some, notorious—professor.* Later, in 1845, Liebig considered becoming professor of chemistry in succession to Daniell but, unfortunately, he was a Lutheran and so, as Bishop Blomfield pointed out, was ineligible for appointment [24]. Furthermore the college was unlucky in that it did not constitute a symbol that could rally the loyalty of a specific group. There was little civic pride in London and the college was not, like University College, the unique achievement of an active political group; it was Anglican, and the Anglicans owed their first loyalty, naturally enough, to the universities. In consequence King's was under-endowed and was later to be desperately pressed for survival.

The existence of two rival colleges in London was, in one respect, unfortunate. The government could not, or felt it could not, make them both universities with powers to award degrees. Nor was King's willing to amalgamate with University College when the latter was avowedly, even aggressively, secular. So, as a compromise, the government created the new-style University of London in 1837; on the foundation of which University College resigned the ambitious title it had previously used ('London University') and adopted its present name. The chief function of the new university was to examine—what else, it was asked, did Oxford and Cambridge universities do?—and, in order to make its examinations as effective as possible it was granted statutory powers to affiliate such colleges as it considered reached sufficiently high standards of tuition and scholarship. The two founder-member colleges were, University and King's, but no limit was set to the number of colleges that could be added to the University. Full use was made of this power. By 1844 no fewer than twenty more colleges

* There was also the unfortunate affair of F. D. Maurice.

had been affiliated in addition to a large number of medical schools in Britain, the University of Malta and the Military Hospital, Ceylon.

While the founding of the University of London was an educational event of the first magnitude its relations with its affiliated institutions were very tenuous; and such relationship as did exist was finally terminated in 1858 when the University of London became a straightforward examination board. It is beyond doubt that this, coupled with the continuing poverty that bedevilled the later history of the two colleges, was thoroughly bad for scientific (as for all intellectual) development in London. Indeed it is remarkable that the colleges, under the circumstances, were able to achieve as much as they did; both University and King's Colleges deserved far better of the country than they got.

On the credit side, the London University, by its system of examination and degrees, enabled a large section of the community to aspire to academic honours that otherwise would have been denied it. At the same time it provided a highly respectable academic umbrella under which the later provincial university colleges—Owens', Liverpool, and Mason's etc.—could grow up. It was felt that the 'open' nature of London degrees prevented any debasement of university degrees such as might have happened had struggling university colleges been tempted to sell their honours in order to fill their lecture halls.

The new university had a most distinguished senate; among its first members being Michael Faraday, Dalton, Nassau Senior, Arnold (of Rugby), Henslow, Beaufort and G. B. Airy. Naturally, therefore, the greatest care was taken to frame its degree requirements and syllabuses in accordance with the very best standards for a liberal education. For the B.A. degree competent knowledge of mathematics, natural philosophy, classics, biology and logic and ethics was required. Having taken the pass degree, the candidate could sit for honours in any of the above subjects or in chemistry or physiology, and beyond this, as opposed to Oxford and Cambridge practice, there was the still more severe examination for the M.A. degree.

Henry Malden, Professor of Greek at University College, a 'gentle and urbane scholar' (H. Hale Bellot), commented on the inclusion of the progressive sciences in the syllabus for the B.A. degree. He observed that, as regards the physical sciences, the ancients were but as children in comparison with the people of his own times. Moreover the sciences were being widely applied and were rapidly changing the conditions of life; this was especially true of the chemical and mechanical sciences. But knowledge of facts was not the main end; balanced and harmonious development was the ideal, and one aptitude was not to be forced above the others. The value of the sciences lay in the habits of observation, accuracy and logical thought that they inculcate [25]. Malden was a man of wide education. Commenting to the Bishop of Durham on the new degree syllabus, he criticised the absence of 'nitric acid' from the requirements in chemistry and suggested the inclusion of the calculus of finite differences in the requirements for mathematical honours [26]. Clearly, he was no narrow specialist.

On the other hand the inclusion of biology in the syllabus troubled John Hoppus, Professor of Philosophy at the same college. Should young men, at an age when their passions must be restrained, 'be introduced to the study of such a subject as "the reproductive functions?" ' . . . surely this would 'inevitably corrupt the youthful mind. It would make all students medical students' [27]. Such was the measure of change, however, that in 1874 T. H. Huxley was employing a lady demonstrator (Miss McConnish) in physiology [28] and, a few years later, in 1879, the London School Board was officially encouraging girls to study the subject [29].

Let us, however, revert to the main topic. For many years the London B.A. examination represented a serious attempt to make the award of a degree conditional upon the candidate's having received an education that was liberal in both senses of the word.* That is to say, an education that is an end in itself and is not so highly specialised that all other considerations are sacrificed to the acquisition of a great and detailed knowledge in one particular subject only. The highly specialised education is usually characteristic of professionalism, in which case the end is not education, but the acquisition of knowledge for the severely practical purpose of earning a living.

The misfortune was that London University was created when no one in England quite knew what a university should do; what its function in society, apart from the obvious one of examination, should be and what should be the nature of its relationship with its constituent colleges.

* A similar educational ideal was propounded by Sir John Herschel. He held that a liberal education should include the physical, biological and social sciences together with ethics [30].

THE UNIVERSITIES

To turn from troubled scenes of lower- and middle-class education to consider that provided for the upper classes is to pass from the pages of Dickens to those of Trollope. Within the walls of the old universities the pace of reform was much more leisurely. Altogether too leisurely for William Cobbett; the sight of the Oxford Colleges aroused him not to voice admiration, but to issue a challenge:* 'Stand forth, ye big-wigged, ye gloriously feeding doctors! . . . Stand forth and face me. . . .'

Not that Cobbett was the only critic, nor were his strictures unjustified. Still the exclusive preserves of the wealthy Anglicans, virtually the only subjects of instruction provided in those great foundations were classics and mathematics. Moreover the internal organisation was such, at both Oxford and Cambridge, that the colleges retained the money and the power whereas the universities were relatively much less wealthy and influential; the colleges, that is, tended to become dead-weights of conservatism. This was anathema to the various radicals and especially to the Edinburgh group. The latter had, in fact, kept up a long range fire from the early years of the century, but their attacks were ineffectual until 1831, when they were able to bring up a really heavy 'piece' in the person of Sir William Hamilton, an Oxonian himself and an able logician. Hamilton possessed a very extensive knowledge of the history of universities and of the requirements imposed by statutes; he was therefore able to point out exactly where and how the English universities had fallen from grace. Crabbed, repetitive and tedious as his writings sometimes are, his command of the undeniable facts and his severely logical exposition made his case a formidable one to answer. More, as he was clearly motivated by the highest ideals and was not calling for radical revolution, he could hardly be rationalised away as a malicious agitator whose attacks could be safely and reasonably disregarded.

Hamilton wanted the control of the universities to be returned to the graduates. He demanded, *inter alia*, that the 'illegal usurpation' by the colleges be ended, that the professoriate be made a reality, and that every M.A. be given his old right to teach (Regency). He insisted that the universities must be national institutions and be effective for their

* *Rural Rides*, 1821.

purpose; and this meant that the domination of the Anglican Church over both the colleges and the universities must end. On the credit side one of the few university institutions that Hamilton found praiseworthy was the examination system.* This was reasonable enough, but on the question of what should be taught in the universities Hamilton was, from the scientific point of view, far less satisfactory. His opinion of the physical sciences was very low: they are, he believed, 'easy and attractive in themselves' but as educational disciplines 'they call out, in the students, the very feeblest effort of thought' [31]. Although these unprogressive opinions on science must have made Hamilton's views more acceptable to conservatives, his polemics took some time to work their effect on the course of university reform; but there can be no doubt of their ultimate influence.

In the meantime the universities had, on their own initiative, managed to achieve some measures of reform. Degree examinations in classics and in mathematics were introduced at Oxford in 1800: for this the credit belongs to Drs. Eveleigh and Jackson. Probably the standard of work, too, was raised and although wealth and rank were accorded deference and privilege for a long time, there occurred at Cambridge at any rate over the whole period a steady decline in the proportion of noblemen and fellow-commoners to pensioners and sizars matriculated. The opening years of the century had been the heyday of the highly privileged but the decline soon set in; the percentages of privileged matriculants over five-year periods being: 1810–14—20.2 per cent; 1815–19—14 per cent; 1820–4—11.5 per cent; 1825–9—9.5 per cent; 1830–4—9 per cent; 1835–9—7 per cent. By the second half of the century the race of blatantly gilded youth was virtually extinct [32].†

At Oxford, an appreciation of the status of mathematical and scientific studies in the university was made by the Rev. Baden Powell, Savilian Professor of Geometry, in the course of a lecture delivered in 1832. Baden Powell lamented that attendance at lectures on chemistry and experimental philosophy had steadily declined, and he was able to show that the proportion of those taking honours in mathematics had, since 1807, also fallen. He believed that the blame lay with the bad

---

* It is fair to add that most of the reforms he advocated were accomplished before the end of the nineteenth century.

† This does not mean that the number of wealthy, or noble, undergraduates was declining. Rather it implies that they went up to the universities more in the spirit of students than of young clubmen.

preliminary education of those entering the university and he called for a wider diffusion of science; maintaining that it was a reproach to send out B.A.s ignorant of science [*33*].

An answer to Baden Powell was almost immediately forthcoming. 'Master of Arts', confessing that he was ignorant of mathematics and therefore incompetent to judge Baden Powell's views on mathematical education, felt called upon to criticise the general tenor of the argument. His own admitted ignorance of science and mathematics made his opinion, that they should occupy no place in general university education, valueless. But it is less easy to refute his main contention that the honours system was an unnatural method of intellectual stimulation; that personal emulation was not proper to a place of learning; and he was entitled to his view that, as regards honours and classes, prizes and medals: 'For my part I think them intrinsically mischievous' [*34*].

This brief debate had occurred at the time of the first visit of the British Association to Oxford; perhaps it was stimulated by that visit. A visit made memorable by Oxford University when it honoured four scientists—Brewster, Faraday, Dalton and Brown—with the award of D.C.L.s. This indicated a glorious disregard for statute and prejudice, for all four were dissenters and would have been unable in the normal course of events to matriculate at the university, much less graduate B.A. But shortly after these happenings Oxford was lost to science for a number of years; the Tractarian movement commenced and men's minds were turned to other matters. Not until after the secession of Newman and his intimate friends and disciples was it possible for the threads to be picked up again. So we must turn our attention to Cambridge, where the atmosphere—thinner, clearer, more mathematical—was perhaps less congenial for the apostolical enthusiasts.

Resembling Oxford in its class-structure and in the fact that the government and teaching staff of the university were predominantly clerical, Cambridge differed markedly from her sister in the strong emphasis that was placed on mathematical studies. Although, as we saw, university mathematics in the eighteenth century was barely in advance of Newton, there sprang up, in the early years of the following century and possibly as a fruition of the work of reformers like Dr. Jebb, a school of very able Cambridge mathematicians. The leaders of this school—Woodhouse, Herschel, Whewell, Maule, Babbage and Peacock—were men of very exceptional ability who, between them, contrived to advance the long neglected study of analytical methods;

founding an Analytical Society for this purpose. With the exception of Maule, who became a judge, these men subsequently rose to positions of academic importance and were able to make major contributions to the progressive reform of mathematical studies. Especially influential in this work were Whewell and Peacock.*

The result of these, and other reforms, was that the Tripos course quickly became, virtually at the beginning of our period, a highly specialised training in mathematics and theoretical physics. Unique in its scientific content, its specialism and its steady historical development, this course of training was also distinguished by its crowning test—the examination—which was, at that time, the great examination not only at Cambridge but also in England.† It was the highest intellectual hurdle in the country and those who acquitted themselves well in it were men of mark; the Senior Wrangler being held in the very highest esteem.

Too much should not be inferred, however, from the high prestige of mathematics and theoretical physics at Cambridge. A few mathematical swallows do not necessarily mean a scientific summer. To understand the function of this excellent and highly scientific education we shall have to take account of the social position and vocational aims of the candidates as well as pay attention to the educational theories and social philosophies of its advocates and critics. Consider first the opinion of Dr. Peacock. Writing in 1841, the Dean of Ely observed that there were two classes of students at Cambridge: (a) the very wealthy who intended to follow no profession, and (b) those who would later enter the professions of law, medicine or divinity. Of the latter divinity was by far the favourite vocation: indeed, at least half the undergraduates intended to take Orders [35]. We may infer that professional mathematicians, even in the form of teachers, were not in great demand. Four years later, in 1845, Dr. Peacock's colleague, the omniscient Dr. Whewell indicated his educational and social philosophies when he remarked that for the upper classes a liberal education was appropriate; for the middle classes there could be an imitation

* In fact, the formal recognition of analytical methods dates from Monday, 13 January 1817, when Peacock distributed the Tripos papers to the candidates: the fourth question on the first paper involved analytical methods. A few years after this, Airy was giving lectures on 'modern' physics.

† The Classics Tripos was introduced in 1822. But before a candidate could sit for this examination he had to be placed as, at least, Junior Optime in the Mathematics Tripos—a requirement that was soon to rankle with the classicists.

thereof, while for the lower orders an elementary education was quite sufficient [*36*]. Developing his ideas, Whewell argued that the liberal education, proper for the upper classes, must be based on mathematics. Newton should be taught in the severe, geometrical form used in *Principia Mathematica*, because premature use of analytical methods, the manipulation of symbols, may well conceal the physical nature of the process under study; that is, the answer may be obtained as from a calculating machine. Together with mathematics and theoretical physics went the study of the classics—the cultural necessity of every educated man—and only after these had been thoroughly mastered could the study of analytical methods be permitted; also the 'progressive' sciences,* as well as the history of science, could be taught.

This austere, yet attractive, programme was justified on the grounds that the study of mathematics was a uniquely valuable mental discipline. It induced 'solid and certain reasoning' and in this way prevented youthful conceits and presumptuousness. In proof of this argument, Whewell drew attention to the number of eminent judges who, in their time, had been high Wranglers. He might also have mentioned the Church, for, between 1800 and 1850, no fewer than 43 men who subsequently became bishops were successful in the Mathematics Tripos.

These opinions of Whewell on the educational value of mathematics were entirely endorsed by Dr. Peacock and by the Rev. B. D. Walsh. The latter, writing in 1837, pointed out that the mathematicians were examined solely in mathematics. This was satisfactory for the study induces habits of industry and application—even if mathematics is later cast aside; essentially mathematical education trains the reasoning powers and is therefore of great value in the training of lawyers [*37*]. In brief, the advocates did not consider the Tripos as a means of producing professional mathematicians. Nor were they blind to the claims of other studies: both Whewell and Walsh urged the introduction of new Triposes in 'progressive' science, etc. Later on Whewell was especially influential in the achievement of these reforms.

This system of intensive, specialised mathematical instruction was not without its critics; and from their writings we can confirm the nature and purpose of the education it represented. But first we can remark that, as early as 1782 Thomas Barnes had discussed the general

* i.e. those sciences which, unlike Newtonian mechanics and astronomy, were 'unfinished' and were therefore progressing towards the goal of theoretical generalisation.

problem of specialised studies and had decided against them. In his view: 'It is a question not only of speculation, but of real importance: "How far is it desirable that a man of learning shall devote himself to one particular object?" ... or ... "Will not the interests of science be best promoted by a more general and extended application to different studies?"' Barnes came to the conclusion that 'The man of "one book" is not likely to understand that book so well as the man of more extended study: there is a general analogy and affinity among all the sciences'. [38] And in 1826 the Edinburgh Review, anticipating Whewell's theory, had rudely dismissed it: 'Of what use the Wrangler? He can reason? But no one reasons so ill as mathematicians!' [39].

Whewell's pamphlet of 1835 (Thoughts on the study of mathematics) was severely criticised by Hamilton, who, in the pages of the Edinburgh Review, denounced the Tripos as '. . . a scheme of discipline more partial and inadequate than any other which the history of education records' [40], and then proceeded to quote 'authority' after 'authority' against the educational value of mathematics. A year later we find Hoppus complaining that university education in England rested on a narrow basis of classics and mathematics to the detriment of the 'progressive' and social sciences. Education should surely be on a broad basis; and was it true that mathematical reasoning was applicable to (say) ethics? [41] In the same year that caustic writer in the British and Foreign Review whom I have already quoted (p. 30) observed that the German universities were disciplined, professional schools, whereas British universities were not, although at Cambridge there is 'a special education of the narrowest kind'.

In 1842 a 'B.A.' wrote that, at Cambridge, after the second term, 'steady, uninterrupted and exclusively mathematical reading' is continued until the final examination, with the exception of the additional studies required for the 'Little-Go' in the second year, but this examination necessitated only about ten days' preparation. It is remarkable that only two subjects are available and that of these only one is usually pursued. As for this subject: the Mathematics Tripos is professional in content and would suit, 'engineers, architects and artillerymen'* but these 'never enter the university. Her sons are

* According to Lyon Playfair, the Duke of Wellington, at a critical time during the Peninsula War, enabled a number of Cambridge mathematicians to obtain commissions in order to overcome an acute shortage of military engineers. This was exceptional and did not, after the war, become an established practice.

clergy and men of leisure.' Judges, it is true, have often taken high honours but the 'legal mind' requires broadening, not narrowing. The mathematician is, at the end of his career, ignorant of all else but mathematics. 'At 22 his education has yet to begin.' The dominance of mathematics is a 'miserable injustice'; other subjects are ignored and real instruction lies with the private tutor. The writer suggested four new Triposes in progressive and social sciences 'open to all but compulsory for none' [42].

But perhaps the most pungent critic of all was A. H. Wratislaw, a classics tutor. He had no hesitation in describing the mathematicians as 'narrow specialists, ignorant of all save their specialism'. The Tripos had no claim to be considered a liberal education: 'It is a professional education of the narrowest kind'. In spite of Whewell there was not even an attempt to encourage the study of the history of science. 'No system could be better adopted calculated to produce professional mathematicians, whose minds were carefully guarded against the intrusion of every non-mathematical thought.' A liberal education is necessary before any display of proficiency in a particular subject is encouraged [43]. Wratislaw's views were much the same as those of the mathematician, Augustus de Morgan. The latter thought that: 'It would not be advisable to permit anyone to obtain a degree on the strength of his acquirements in one branch of knowledge only' [44]. And, on another occasion, de Morgan advocated a degree course that 'preserves this one characteristic; namely the cultivation of more than one capability, the creation of strength in more than one way of using intellect' [45].

SUMMARY

The Tripos was a highly specialised examination, implying highly specialised study; and, although 'Little-Go' demanded some classics and religious knowledge it was unlikely that a budding Wrangler would be 'plucked' (as the Victorians had it) for deficiency in those subjects [46]. But there was nothing professional about the Tripos; it was defended on the grounds of its importance for a liberal education; it was criticised on the grounds that it was not liberal in content; and, with the possible exception of Walsh, no one seems to have favoured the study of mathematics to the exclusion of other subjects.

Professional studies, even at that late date, had no place in the university. Law was studied at the Inns of Court, medicine at the

London hospitals and for the clergy no special training was thought necessary. The Victorian engineer and his predecessors were trained in the old craft apprenticeship system (e.g. Bramah, Maudslay, Naismyth, Brunel, Bessemer and Whitworth). The universities were concerned with the liberal education of men of a privileged class who would later adopt suitable professions or else follow a life of leisure. The educational ideal was the Christian gentleman; if he was a scholar, then so much the better; if not, then he would benefit from the corporate life in the university.

Against the critics must be set the important fact that the young Wrangler, on leaving the university did not enter a highly specialised social milieu. The Tripos, in fact, provided the means whereby brilliant young men could secure early recognition of their talents; the young lawyer or clergyman who had been a high Wrangler was a notable person with improved chances of promotion or preferment in his chosen career. Also, the educational theory that public men: judges, bishops, statesmen, should be well grounded in the principles of Newtonian science was, surely, an excellent one?

In one respect only does the theory of the Tripos appear to be ill found. The defect of the system lay in its virtues. With our experience of a much more complex society and with our sophisticated psychology we are well aware that the pursuit of mathematics requires special tastes and aptitudes; qualities of mind rarer, perhaps, than those associated with history, biology, chemistry, literature, etc., but not for that reason to be regarded as necessarily higher. Expressly intended as a liberal education, the Tripos could not but attract the able mathematician and, in this way, other things being equal, the non-mathematical and the mathematical student were encouraged to compete in mathematics; a competition that could hardly be fair. It is strange that none of the critics of mathematical education made the obvious protest that mathematicians are born, not made. We can, of course, explain this absence of criticism by reference to contemporary ignorance of psychology but such an explanation is not complete; it can be complete if we recall that there was not, at that time, a class of professional mathematicians or scientists. After all, it has never been the custom to put the 'amateur' and the 'professional' in mutual competition.

Correlatively, the university was not conceived to be, *inter alia*, a centre of research with the purpose of advancing knowledge: science— natural philosophy, natural history, chemistry, etc.—was still largely conducted by amateurs. Only the far-sighted could have discerned the signs of coming change of which one was the marked rise in the quality of university staffs, men like Whewell, Peacock, Airy, Stokes, Greene, Baden Powell, Sedgwick at Oxford and Cambridge; de Morgan, Graham, Lyell, Daniell, Wheatstone, at the London colleges. And there was the inclusion of medical studies within the walls of the University of London.

THE REFORM OF SOCIETIES AND THE STATE OF SCIENCE

The early 1830s were years of reform not only in politics but in many other fields as well. Scientific organisations, particularly the Royal Society, were among those institutions to shiver in the cold winds of criticism. A scientific society composed largely of men ignorant of science was an inviting target for critics like Sir James South (the amateur astronomer) [47], Charles Babbage, A. B. Granville, Col. Everest [48] and others.

A positive reaction to the dilettantist state of the Royal Society was the formation, in 1831, of the British Association. This, like the foundation of University College, was partly inspired by a German initiative. In 1822 Lorenz Oken, the metabiologist, had succeeded in organising, in Leipzig, a meeting open to those engaged in medicine and natural science. Oken was at that time a political refugee and so suspicious were the German despots that only thirty-two philosophers could summon up courage to attend this, the first philosophical congress in Europe. In the following years meetings were held in Halle, Würzburg, Frankfurt and Dresden. The conference evidently gained in respectability for the Emperor Maximilian was patron of the 1827 meeting at Munich and, in Berlin in 1828, von Humboldt was President and the King of Prussia a guest. Another important visitor on this occasion was Charles Babbage, then travelling on the Continent.

Babbage, whose part in the reform of Cambridge and therefore of English mathematics we have already noticed, was a man of near universal genius. A Lucasian Professor at Cambridge who, for eleven years, scandalously neglected his teaching duties but who, for the rest of his life, devoted his time and his fortune to science, he seems to belong to the eighteenth rather than to the nineteenth century. On the other hand, as the author of the forward looking *On the Economy of Machinery and Manufactures* and as the pioneer of the computer he belongs, in spirit, to the second half of the twentieth century. His first idea for a computer had been the 'difference engine' on which he began

work in 1823; this was followed by a plan for a much more advanced
'analytical engine' which made use of the punched-card technique made
familiar by the Jacquard loom. But Babbage's ideas had run far ahead
of his—or anyone else's—resources: the difference engine proved very
difficult and in 1827 his health broke down. On medical advice he
therefore spent a year travelling on the Continent; one consequence of
this tour was the book we have just mentioned [49], another followed
from his visit to Berlin.

A man like Babbage could hardly fail to be impressed by the Berlin
assembly; and, on his return home he wrote an enthusiastic description
of what he had witnessed for Brewster's *Edinburgh Journal* [50]. This
article received wide publicity and was much discussed in British
scientific circles. In the following year Robert Brown, the botanist and
discoverer of 'Brownian Motion', attended the Heidelberg assembly.
He, too, gave a favourable report on the German enterprise, and he in
turn was followed by other commentators who extolled the German
society.

While these German developments were in progress there was in
England a widespread feeling that English science was in decline.
(This seems very curious when we reflect that the forty years following
the death of Davy saw the great, and in many ways unparalleled
achievements of Faraday, Darwin, Joule, Lyell, Kelvin, Maxwell and
many others.) Davy had intended to write a book on the decline of
British science, but died before it could be completed. Sir John
Herschel was another authority who believed that the British were
dropping behind, especially in mathematics and chemistry. Certainly
Babbage fully shared the general unease, for in 1830 he brought out
his celebrated work on the state of science in England [51]; a book
which aroused great interest and was evidently partly inspired by his
experiences of the *Deutsches Naturforschers Versammlung* and other
Continental societies. This famous work was largely an attack on the
Royal Society but the State and society at large did not escape chastise-
ment: science, he remarked, received scant recognition in England
whereas in France and Prussia men of science were awarded high
honours and sometimes even reached ministerial rank (e.g. Laplace,
Chaptal). Our schools and universities did not encourage science and,
in fact, 'The pursuit of science in England does not constitute a
distinct profession as it does in many other countries'.

A long and favourable review of this book appeared in the *Quarterly
Review*. At the end of the article the writer, Sir David Brewster,

advocated the formation of an organisation, similar to the German one, to forward the interests of British science; he added an appeal to the 'nobility, gentry, clergy and philosophers' to lend their aid in achieving this [52]. The appeal was not ill-timed for it contributed directly to the formation of the British Association for the Advancement of Science* (York 1831); a consequence of the 'decline of science' agitation and of the writings of Babbage and Brewster in particular.

The debt which the reform movements in science and education owed to the Edinburgh group was very great. Brewster, in particular, had long been warning the country of the consequences of the neglect of science; denouncing governmental ignorance, the bad organisation of scientific boards and institutions and what he thought were very unjust patent laws [53]. This pessimism was not, of course, confined to the physical scientists: William Swainson, a student of natural history, compared the national patronage of science in other countries with its neglect by the British Government. We gather from Swainson that every country—France, Italy, Germany, even Russia—encouraged its scientists; but in England the sole criterion was that of marketable value [54].

In the light of these and other charges of the neglect of science [55], it is only fair to see what defence can be made of British practice. One of the first to speak on behalf of British science was J. F. Daniell who, in the course of his introductory lecture from the chair of chemistry at King's College, took the opportunity to refute some of the allegations made [56]. Daniell spoke with authority, for he was a chemist of recognised eminence. Another defender was Dr. A. B. Granville who, when engaged in criticising the Royal Society, pointed out that, were the level of scientific activity a function of the social institutions enumerated by the critics, it is quite impossible to see how science could have advanced in the past, still less how it could ever have commenced. For indisputably the social incentives to science were less in the past than they were in Granville's time and the obstacles were greater.

Professor Moll, of Utrecht, took the liberty of defending British science [57]. There was no lack of talent in the country: Herschel, Babbage, Brewster, South, Faraday, Dalton and many other names were to its credit. Furthermore in France science was not free and, in

* It was not without interest that the men who founded the Association referred to themselves as 'cultivators' of science. The expression originated from a mis-translation of the word 'naturforschers'.

any case, political patronage was a very doubtful proposition. The British men of science are free, versatile and not over-specialised, and this is all to the good for, as Cicero has it: 'Omnes artes quae ad humanitatem pertinent, habent quoddam commune vinculum, et quasi cognatione quadam inter se continentur'. For this defence of British science Moll and his prompter Faraday brought upon themselves the critical thunder of Brewster [*58*].

Whatever the true state of British science in the 1830s, it cannot be doubted that the invective of the critics coupled with the positive action they took—the founding of the British Association—was of immense benefit to science. The prophets thus nullified their own prophecies. But we still have to face the question: what reasons can be found to account for the debate? Very obviously there was, relatively speaking, no lack of first class talent, even genius, in the country. However, the phrase 'Cultivators of Science' reveals the temper in which science was pursued in England, and it would be fair to say that there was a marked absence of scientists of the second and third rank. The judgments were comparative and quantitative. In France and the German States there was greater scientific activity than in England: more papers were being published and, possibly, more new fields were being opened up; also there were more scientists in the lower ranks. Scientists of the second and third rank may well be professional and/or applied scientists (certainly this is the case to-day). Throughout the nineteenth century first France and later Germany led the world in the business of discovering how to apply science. In Britain such second and third rank scientists as there were, were to be found in such professions as the law, the armed forces (Kater, Sabine, Ommaney, Beaufort, etc.), the church and medicine; that is, they were to a large extent amateurs.

On one point, at least, the critics had right on their side. In the question of educational institutions there was little to be said on Britain's behalf. The neglect of science will form a curious story for aftertimes, thought Augustus de Morgan, for 'among the first commercial peoples of the world, who depended for their political greatness on trade and manufactures, there was not, generally speaking, in the education of their youth one atom of information on the products of the earth, whether animal, vegetable or mineral, nor any account of the principles whether of mechanics or of chemistry which, when applied to those products, constituted the greatness of their country' [*59*]. In the year 1832 Baden Powell, watching the spread of the mechanics' institutes movement, warned the upper class that unless they paid attention to science 'the consequence must be that that class will not long remain the HIGHER'.*

Ten years later W. Cooke-Taylor, travelling through Lancashire, noted that, individually, no men do more for science than do Lancastrians, but there was a want of organisation and this was a national calamity. 'Scientific knowledge has ceased to be a luxury; it has become a necessity of life' but 'the manufacturing youth of the higher and middle classes are not trained for the order to which they must eventually belong' [60].

The Rev. James Booth, LL.D., F.R.S., the Principal of the Liverpool Collegiate Institution, and another of the rather formidable clergymen concerned with nineteenth-century education and science, proclaimed in 1846 that, while the education of the poor was just beginning, that of the upper and middle classes was retrograde. A senile apathy opposed the spread of sound and practical learning. The universities excluded science when our national existence depended on it; even in the mechanics' institutes there was more amusement than hard work. Despite the establishment of such colleges as those at Putney (engineering) and Cirencester (agriculture) the manufacturing districts were still without a technical college. While in England scientific education was unrecognised, all Europe was pressing forward to outstrip her [61]. Booth, whom we shall meet again, was a vigorous and clear-thinking man with some very decided opinions; his observations on European competition were very apposite: the German Technical High Schools were being rapidly developed at this time and, of the Switzerland of 1836, Dr. John Bowring had reported a country of universal education with reasonable technical instruction and a good school of manufactures at Geneva [62]. One cannot accuse Booth of undue alarmism.

It is self-evident that the diffusion of scientific knowledge is a function of the educational system and it was here that the weak point lay, as these men had seen. To call attention to this the Central Society of Education was founded in 1837 by a distinguished group that included Lord Denman, de Morgan, Benjamin Heywood, Lord John Russell, Neill Arnott and Sir John Lubbock. In their first publication they stated bluntly that the main defect of English education, 'to which most of its injurious and inefficient working may be traced, is the total want of national organisation' [63]. A want of organisation, it may be

* His capitals.

added, that such voluntary enterprises as Brougham's Society for the Diffusion of Useful Knowledge, the *Penny Cyclopaedia*, etc. and even, in the long run, the mechanics' institutes were unable to offset.

A cause and a justification for the pessimism and an effect of the retrograde educational system was that the trek of young chemists to the German universities had begun well before 1850. Among the first to go were W. C. Henry, A. W. Williamson, Edward Frankland, Lyon Playfair, Edward Turner, William Gregory, the Muspratt brothers,* J. H. Gilbert,† and Warren de la Rue. The favourite teachers were Liebig, at Giessen, and Bunsen, at Heidelberg. Before the end of the century it became standard practice for chemists—and for many other scientists as well—to go to Germany to obtain their higher education.

An important reason for this practice, at least in the early days, has recently been made clear. J. B. Morell has called attention to an important but little known pamphlet published by William Gregory in 1842 [*64*]. This contrasted the extraordinarily parsimonious self-help system under which the Scottish professor laboured (see p. 33) with the much more generous treatment that his German contemporary received. Liebig, for example, was provided with a good salary, a free house, very adequate running expenses, a salaried assistant and a salaried laboratory steward. Accordingly full-time study for one academic year at Giessen cost only £6. 14. 0d: nothing like this could be had anywhere in Britain. Moreover, as W. V. Farrar points out, a Giessen student could obtain the coveted Ph.D. degree while there was no such academic incentive to research provided by any British university [*65*].

As a footnote we may quote the opinion of Liebig on British science. He had first visited England in 1837 and when he came again, in 1842, his visit was rather like a triumphal procession; he was accompanied on his tour by the Duke of Westminster and by his former pupil, Lyon Playfair, now rising rapidly in the world of affairs. Liebig observed that England was not the land of science, for '. . . only those works which have a practical tendency awake attention and command respect, while the purely scientific works, which possess far greater merits, are almost unknown. . . . In Germany it is quite the contrary.'‡

There was a good deal of truth in this. But things are rarely as simple as they seem: had Liebig been in Manchester a few months before he

* Of Widnes chemical industries.
† Of Rothamsted.
‡ Letter to Faraday, 19 December, 1844.

wrote his letter he might have revised his opinion. When Dalton died the body of the great scientist lay in state in the town hall and some forty thousand people passed the bier to pay their last respects. The funeral procession was nearly a mile long and included representatives of most of the public bodies of the town. In fact, 'the town was occupied for a time with the funeral of Dalton; business ceased; the streets were thronged with numberless spectators; and the police of Manchester attended with a badge of mourning' [*66*].

We must remember that, in the nineteenth century, the scientific life of England was not confined to London and Cambridge. Round the Manchester Literary and Philosophical Society a distinguished and very creative school of scientists had grown up with Dalton as its inspiration. This school established Manchester as one of the centres of European science; a centre that had closer relations with Scotland and, later, Germany than it did with London and Cambridge. Following Dalton, Manchester could boast of such men as Joule, Eaton Hodgkinson, Balfour Stewart, Osborne Reynolds and Ernest Rutherford [*67*].

REFERENCES

[*1*] For an account of the problems (and their solutions) of applying steam power to the textile industries see R. L. Hills, *Power in the Industrial Revolution* (Manchester University Press, 1970).

[*2*] Thomas P. Hughes, Introduction to *Selections from the Lives of the Engineers*, by Samuel Smiles (M.I.T. Press, Cambridge, Mass. 1966).

[*3*] W. H. G. Armytage, *The Rise of the Technocrats: a Social History* (Kegan Paul, London 1965) p. 41 et seq.

[*4*] Lord Ernle, *English Farming Past and Present* (Longmans Green, London 1936) pp. 196, 216–7.

[*5*] See Cobbett's *Parliamentary Debates* (1807). J. L. and Barbara Hammond, *The Rise of Modern Industry* (9th edition, Methuen, London 1966) p. 232; *The Town Labourer, 1766–1832* (Longmans Green, London 1966) pp. 65–6. For details of Gilbert's career see A. C. Todd, *Beyond the Blaze: a Biography of Davies Gilbert* (D. Bradford Barton, Truro, 1967).

[*6*] Patrick Colquhoun, *A Treatise on Indigence* (London 1806) pp. 148–9.

[*7*] Henry Brougham, *The Life and Times of Henry, Lord Brougham* (3 vols., Edinburgh and London 1871) i, pp. 71–2.

[8] Mona Wilson, *Jane Austen and Some Contemporaries* (Cresset Press, London 1938). See also *D.N.B.* and Elizabeth C. Patterson, 'Mary Somerville', *B.J.H.S.* (December 1969) iv, p. 311.

[9] J. W. Hudson, *The History of Adult Education* (London 1851).

[10] 'Country Gentleman', 'The Consequences of Scientific Education', reviewed in *Edinburgh Review* (1827) xlv, pp. 189–99.

[11] E. W. Grinfield, 'A Reply to Brougham's "Practical Observations on the Education of the People" ' (1825), reviewed in *Edinburgh Review* (1825) xlii, p. 206.

[12] *Quarterly Review* (1825) xxxii, pp. 410–28.

[13] John Penfold Thomas, 'Reports of Two Speeches Against the Establishment of Mechanics' Institutes at Rotherhithe and Southwark' (1825–6). Privately printed.

[14] Mabel Tylecote, *The Mechanics' Institutes of Lancashire and Yorkshire before 1851* (Manchester University Press, 1957), pp. 26–52.

[15] C. Delisle Burns, *A History of Birkbeck College* (University of London Press, 1923).

[16] George Birkbeck, speech at public meeting, 11 November 1823 (London 1823).

[17] Benjamin Heywood, 'Address to the Mechanics, Artisans etc., delivered at the Opening of the Manchester Mechanics' Institution' (Manchester 1825).

[18] Henry Brougham, *Practical Observations on the Education of the People* (London 1825).

[19] Rowland Detrosier, 'An Address on the Advantages of the Intended Mechanics' Hall of Science', 31 December, 1831 (Manchester 1832).

[20] J. G. Goddard, *George Birkbeck* (London 1884) p. 208. The most recent biography of Birkbeck is by Thomas Kelly, *George Birkbeck: Pioneer of Adult Education* (Liverpool University Press, 1957).

[21] 'Christianus' (Rev. George D'Oyly), 'A Letter to the Rt. Honourable Robert Peel on the Subject of the London University' (London 1828).

[22] 'Second Statement by the Council of the University of London' (1828). Bound in *Pamphlets Relating to the University of London*, i, in University of London Library.

[23] 'Statement by the Council of the University of London Explanatory of the Nature and Objects of the Institution' (1827), op. cit. [22].

[24] Timothy Holmes, *The Life of Sir Benjamin Collins Brodie* (London 1898] quotes a letter from Blomfield to Brodie, Sen., 6 May, 1845.

[25] Henry Malden, 'On the Introduction of the Natural Sciences into General Education', a lecture at U.C.L., 15 October, 1845.

[*26*] Henry Malden, letter to the Bishop of Durham, 1 May, 1845, in University of London Library.

[*27*] John Hoppus, letter to the Bishop of Durham, 1 May, 1845, in University of London Library.

[*28*] Bernard Becker, *Scientific London* (London 1874), p. 182.

[*29*] Quoted by D. M. Turner in her *History of Science Teaching in England* (Chapman and Hall, London 1827).

[*30*] Sir John Herschel, *London and Edinburgh Philosophical Magazine* (1836) viii, pp. 432–8.

[*31*] Sir William Hamilton, *Collected Essays* (London 1853) p. 780 et seq.

[*32*] D. A. Winstanley, *Early Victorian Cambridge* (Cambridge University Press, 1940) p. 415.

[*33*] Rev. Baden Powell, 'The Present State and Future Prospects of Mathematical and Physical Studies in the University of Oxford', a lecture delivered in 1832. Bound in the Augustus de Morgan papers, University of London Library.

[*34*] 'Master of Arts', 'A Short Criticism of Baden Powell's Lecture' (1832) ibid.

[*35*] Rev. George Peacock, *Observations on the Statutes of the University of Cambridge* (London 1841) p. 151 et seq.

[*36*] Rev. William Whewell, *Of a Liberal Education* (London 1845). His social philosophy is set out at the beginning of the book. See also his *Thoughts on the Study of Mathematics as Part of a Liberal Education* (Cambridge 1835).

[*37*] Rev. B. D. Walsh, *Historical Account of the University of Cambridge and its Colleges* (Cambridge 1837) p. 86.

[*38*] Thomas Barnes, *Manchester Memoirs* (1785) i, loc. cit.

[*39*] *Edinburgh Review* (1826) xliii, pp. 315–41.

[*40*] *Edinburgh Review* (1836) lxii, pp. 315–41.

[*41*] John Hoppus, *Thoughts on Academical Organisation and Degrees in Arts* (London 1837) pp. 22, 37.

[*42*] 'B.A.', *University Education* (London 1842) p. 17 et seq.

[*43*] A. H. Wratislaw, *Observations on the Cambridge system. Partly in reply to, partly suggested by, Dr. Whewell's educational publications,* (Cambridge 1850) p. 16 et seq.

[*44*] Augustus de Morgan, 'State of the Mathematical and Physical Sciences in the University of Oxford', *Quarterly Journal of Education* (1832) iv, p. 191.

[*45*] Augustus de Morgan, 'Thoughts Suggested by the Establishment of the University of London', a lecture delivered in 1837. Bound in the de Morgan papers in the University of London Library.

[*46*] Silvanus P. Thompson, *The Life of Lord Kelvin*, (2 vols., London

1910) i, p. 23 et seq. See also W. W. Rouse Ball, *A History of the Study of Mathematics at Cambridge* (Cambridge 1889) p. 214

[47] Sir James South, *Charges against the President and Councils of the Royal Society* (London 1830).

[48] Lt. Col. Everest (A series of) *Letters to H.R.H. the Duke of Sussex remonstrating against the conduct of the learned body*, the Royal Society (London 1839).

[49] Richard Rosenbloom, 'Men and Machines', *Technology and Culture* (1964) v, p. 489.

[50] Charles Babbage, 'Account of the Great Congress of Philosophers at Berlin', *Edinburgh Journal of Science* (1829) x, p. 225. See also his *Passages from the Life of a philosopher* (London 1864).

[51] Charles Babbage, *Reflections on the Decline of Science in England* (London 1830).

[52] David Brewster, *Quarterly Review* (1830) xliii, p. 305. See also O. J. R. Howarth, *The British Association: a Retrospect, 1831–1931* (The British Association, London 1931).

[53] David Brewster, 'Observations on the Decline of Science in England', *Edinburgh Journal of Science* (1831) n.s. v, p. 1.

[54] William Swainson, *Preliminary Discourse on the Study of Natural History* (London 1834) p. 348.

[55] G. A. Foote, 'The Place of Science in the British Reform Movement, 1830–1850', *Isis* (1951) xlii, pp. 192–208. See also J. B. Morrell's important paper, 'Individualism and the Structure of British Science in 1830', *Historical Studies in the Physical Sciences*, edited by Russell McCormmach, iii (University of Pennsylvania Press, 1972).

[56] J. F. Daniell, 11 October, 1831 (London 1831).

[57] 'Foreigner' (Professor Moll), *On the Alleged Decline of Science in England* (London 1831). See also Nathan Reingold, 'Babbage and Moll on the State of Science in Great Britain', *B.J.H.S.* (June, 1968) iv, p. 58.

[58] David Brewster, *Edinburgh Journal of Science* (1831) n.s., v, p. 334.

[59] Augustus de Morgan, *Quarterly Journal of Education* (1832) iii, p. 72.

[60] W. Cooke-Taylor, *Notes of a Tour in the Manufacturing Districts of Lancashire* (London 1942) pp. 279–81.

[61] Rev. James Booth, *Education and Educational Institutions Considered with Reference to the Industrial Professions and the Present Aspect of Society* (London 1846).

[62] John Bowring, *Report on the commerce and manufactures of Switzerland* (A Command Paper, London 1836).

[63] *Central Society of Education: First Publication* (London 1837).

[64] J. B. Morell, *Ambix* (July 1969) xvi, loc. cit.

[65] Ibid.

[*66*] Robert Angus Smith, 'Memoir of John Dalton and History of the Atomic Theory up to His Time', *Manchester Memoirs* (1856) xiii, pp. 1–298.

[*67*] *John Dalton and the Progress of Science*, edited by D. S. L. Cardwell (Manchester University Press, 1968) pp. 1–9.

CHAPTER FOUR

The Mid-Century:
1840–70

In his discussion of bureaucracy, Max Weber notes that the modern development of the bureaucratic system brings to the fore the practice of rational, specialised and expert examinations. Whereas previously the 'cultivated' man had been the ideal, under the impact of bureaucracy it becomes the 'expert'. With this movement are associated social-levelling tendencies and the rise of democracy, together with the separation of the worker from the ownership of the means of production, and that limitation of function characteristic of bureaucracy. In this context it was of no little significance that the one feature of the English universities of which the reforming radicals thoroughly approved was their examination system.

So far, as has been shown, the educational ideals of the upper and middle classes of English society have certainly been 'liberal'. Even the system proposed for the working classes had a strong 'liberal' element (see above p. 43). Furthermore, English science had continued to be, to a large extent, the preserve of amateurs and had only been loosely linked with technology. The amateur necessarily owns the 'means of production' although such may be both large and costly—as for example, was the Earl of Rosse's giant reflecting telescope at Parsontown, in Ireland.

It can, in fact, be said that up to the mid-century the social organisation of science in England had not yet commenced. In essence, therefore, what has been said above is no more than a preface; important and not without historical interest, but still a preface. We have seen that scientific societies were to a large extent, composed of men interested in science but not necessarily active therein; and we have inferred that the 'Decline of Science' movement was actuated by the recognition that Continental nations had begun to organise science for the national good: that it was taught in their schools and universities, that it was beginning to permeate their workshops and factories and that the advice of men of science was sought and heeded by their governments.

OF MECHANICS' INSTITUTES, TECHNOLOGY
AND OTHER MATTERS

Less than ten years after the great events of 1823–5 it began to be quite clear that all was not well with the institutes. The trouble may be summed up in the words of Dr. Hudson: 'it was proved, upon un-doubted testimony, that these societies had failed to attract the class for whom they were intended, by their founders, to benefit'.* That is, as Brougham had noticed in 1835, the mechanics and artisans—the élite of the working class—were abandoning the institutes and their places were being usurped by clerks and shopmen. The lecture courses, started with so much enthusiasm, were falling off badly and the laboratories and workshops were being used less and less.

There were ample facts and figures to bear this out. For example, between 1839 and 1852 the Leeds institute increased its membership by nearly tenfold but the average mathematics class declined from 36 to 24 while the chemistry class fell from 19 to 13. At Manchester the classes in physical science dropped from 235 in 1835–9 to 127 in 1840–4, and to 88 in 1845–9. Side by side with this went an increase in the numbers of lectures on light literature and the drama and even on quite bogus sciences—phrenology and mesmerism were very popular.

This raises the question: why did the institutes fail in their primary duty? It is not difficult to propound the answers.

In the first place there was the illiteracy of the classes for whom they were intended. For this, the primary cause, the defective education of the country was to blame and beyond that were causes which are not immediately relevant to this account.

As Thomas Coates of the S.D.U.K. remarked [1], it was unreal to exclude economics and politics from the syllabus, yet to include atomic theory and astronomy. The former subjects were of vital interest to the intelligent artisan: to expect him to forgo them in favour of abstract science was to demand from him a philosophic detachment on a truly heroic scale.

While it is quite true that a few men—James Young, founder of the paraffin industry, John Mercer of Oakenshaw and Peter Spence the chemist-industrialist, could and did use these science classes to climb to the top of the tree, they were very exceptional men who would probably have risen under any circumstances. For the common run of folk things were different. Unless the institutes could provide some

* J. W. Hudson, *The History of Adult Education.*

kind of diploma, to be awarded on completion of an approved course in an approved manner, it is difficult to see what objective incentive there could be to protracted and systematic study. While a few of the largest institutes could do this (notably the Edinburgh School of Arts, which first awarded diplomas in 1835), it was still necessary that such diplomas should be financially worthwhile; that is, that employers generally should recognise their value and give preference and promotion to the holders of them.

The middle-class invasion is a fact beyond dispute* and it was some-times suggested that resulting class-consciousness had the effect of driving the workers away from the institutes. This, however, was clearly secondary and merely an effect of the major causes enumerated above; although this is not to deny that class feeling had some effect. It also demonstrates the deficiencies of middle-class education and, in fact, of secondary education generally—that the institutes were used to a large extent by the wrong classes does not mean that they were useless. T. E. Cliffe-Leslie [2], examining the situation in 1852 dis-covered that the institutes were used by the lower middle-classes in lieu of free public libraries. This was not surprising for while there were in 1847, 107 public libraries in France, 44 in Prussia, 48 in Austria, 14 in Belgium, over 100 in the U.S.A., there was only one in England. Being self-supporting, the institutes had no option but to cater for public tastes; consequently fiction, periodicals and light literature generally came to figure prominently on the library shelves.

Whatever virtue may have inhered in the legal maxim that the Queen could do no wrong, the belief of many of Victoria's subjects that the State could do no right was of more doubtful validity. The belief in individualism, the frequently expressed dislike of State interference and of centralised administration, were responsible for many of the educational ills of the nineteenth century. Even Hudson disapproved of State interference: 'On the whole the experience of the past is proof of the danger of government interference, and of the instability of extreme centralisation, while it affords conclusive evidence of the superior and enduring value of voluntary efforts'.† He had in mind the failure of some fifteen Irish institutes that had enjoyed State support but had failed, and also the ill-success that had attended the government sponsored schools of design started in 1837. There were plenty to agree with Hudson—for example, Herbert Spencer (*State Education*

* *Vide*, J. W. Hudson, Thomas Coates, Brougham, etc.
† Op. cit., p. 188.

Self-Defeating, 1851),* Thomas Wrigley (*A Plan whereby the Education of the People may be secured without State Interference*, 1857) and the Dean of Hereford (*On Self-Supporting Education*, 1853). To justify the individualist point of view it was necessary merely to remark that whereas free enterprise had founded all the mechanics' institutes as well as the noble institutions of University College and King's College,† endowments, as Adam Smith had proved and no one had since doubted, had produced the educational sterility of Eton School and the unreformed universities. And if that did not satisfy you, were you prepared to hand over education to the tender mercies of the fox-hunting, port-swilling squirearchy? The individualist argument, had therefore, a respectable kernel of truth; although at the same time J. S. Mill had come to the opposite conclusion, maintaining that 'Education . . . is one of those things which it is admissable in principle that a government should provide for the people'.‡

Be this as it may, when the social changes within the institutes were noticed, there were vigorous attempts at reform, improvement and reorganisation. When the full significance of illiteracy was recognised, the leading spirits of the movement made determined efforts to overcome it. For the simpler folk Lyceums were established where they could acquire the three 'Rs' without the humiliation of 'going to school' but, on the other hand, without being expected to jump the fence to atomic theory. Also, acting on a very sound principle, the larger institutes, notably those at Manchester and Liverpool, established day schools for boys and girls. But perhaps most important of all was the invention of Union; the first example of which—the Yorkshire Union of Mechanics' Institutes—had been originally proposed by Edward Baines in a letter published in the *Leeds Mercury* in September 1837.

The purposes—and advantages—of federation were obvious: they were to enable a rational system of lectures to be created and the free circulation of books and pamphlets to be started. The lectures proposed were to have included systematic courses in mechanics, chemistry, economics and statistics; it was also hoped that a diploma of merit could be awarded to the industrious, and that this would achieve general recognition from employers. Unfortunately, for the first few

* From *Social statics*.
† Durham University, so closely modelled on Oxford and Cambridge, was less praiseworthy from the radical and individualist points of view.
‡ *Principles of political economy* (London, 1848).

years little was accomplished apart from a few lectures on applied chemistry and mechanics, and it proved impossible to obtain a permanent lecturer in the physical sciences as had been intended; instead the lecture courses were given by part-time and voluntary workers. Not until 1846, when an efficient secretary and agent was engaged, did the Union really begin to expand. By mid-century it embraced over 100 institutes with a membership of nearly 20,000 individuals and was, at that time, the largest educational organisation in the country. Similar unions were also founded in London, Lancashire, Cheshire, Scotland, the Midlands and the North.

It would not be claiming too much to suggest that from the day when the Yorkshire Union was founded we can date the beginning of the effective social organisation of science in the country. This was the first organised attempt, on anything approaching a national scale, systematically to teach and to forward the 'pure' and applied sciences. The first institutes had been launched with the hope that they would be carried along by the stream of progress; the fundamental need for organisation had not been forseen.

The report of the 1851 census noted the location of the individual institutes* and the nature of the classes they held [3]. The greatest concentrations of science classes were in South Lancashire and the West Riding, thus confirming Cooke-Taylor's observations. Other notable areas were Tyneside, the ports of Southampton and Devonport and the Cornish mining districts. Conversely the backwardness of the Black Country had continued since Brougham had first noticed it in 1825. This was possibly due to the dispersed nature of Black Country industry: a pattern of small masters would not be favourable for such collective efforts as the founding of an institute would require. Also wealth was more dispersed in such a society and there was less chance that a wealthy and benevolent person would either give a lead or make a large donation.

In order to give some proportion to the picture, however, it must be remembered that there were, at that time, only 700 institutes in the country with a total membership of about 110,000. That is to say, considerably less than one per cent of the population were members of institutes. Even in an advanced county like Yorkshire, only two per cent of the people were members, and in the leading city of all—Manchester—it was estimated that less than one third of the mechanics belonged to an institute. This, it seems, constituted a serious indictment

* See p. 75.

of the social conditions of the wealthiest country in the world. In the manner of the iceberg, only a small part of the body social was open to survey, the greater part being submerged and out of sight. The analogy is not inapt for it was widely believed that the 'laws' of political economy were as immutable as those of hydrostatics.

Distribution of Science Lectures and Classes in
1851 (Mechanics' Institutes, etc).

THE SOCIETY OF ARTS

The Society for the Encouragement of Arts, Manufactures and Commerce (now the Royal Society of Arts), was founded in 1754. There are few, if any, organisations in the United Kingdom whose records of services freely given can rival that of this admirable Society. Agriculture, industry, applied art, education and 'pure' and applied science have all been benefited and the rooms of the society of Arts served as incubator and clearing house for many of the most constructive ideas for the development of science, technology and social reform in the nineteenth century. During the Presidency of the scholarly and liberal Prince Consort, the active members included such interesting and dominating personalities as Henry Cole, Lyon Playfair, Edwin Chadwick, J. F. D. Donnelly, R.E., as well as many others whose services are too well known to require description.

From the time of the revolution onwards the French had held national exhibitions at more or less regular intervals and the first idea of an international exhibition seems to have occurred to M. Buffet, Minister of Commerce, in 1848. Nothing came of this original conception for French manufacturers appeared reluctant to sponsor it. However, the idea travelled to England where the soil proved to be more fertile; there it aroused the interest of Henry Cole and a small group of members of the Society. From their subsequent initiative was born the deservedly famous Great Exhibition of 1851. In the execution of the project Henry Cole, to whom the chief credit is due, was joined by Lyon Playfair and the whole enterprise had the backing of the Prince Consort.

Certain modern writers have expressed the opinion that the Exhibition was not important; but this is a specialist view which cannot be maintained. An assessment of its direct and indirect effects will show clearly how important it was—at least, for science and technology. On the one hand it represented the concrete embodiment of the idea of progress in so obvious a fashion as to penetrate the general consciousness of the community. As J. B. Bury wrote, 'it was in one of its aspects a public recognition of the material progress of the age and the growing power of man over the physical world' [4]. Hence it resulted in an enhanced social approval of science and technology and it stimulated the advocates of science to renewed efforts.

Also it created a fashion for international exhibitions: during the second half of the century exhibitions were held in many cities all over the world; by 1872 they had penetrated to places as remote as Santiago, Lima and Kioto. Whether these contributed to international amity is a debatable question; what appears to be less doubtful is that they served as milestones of progress: the Victorian could note, exhibition by exhibition, the advances made in the various sciences, technologies and applied arts both by his own country and by its trade rivals. Accordingly he could feel satisfied or he could take steps to remedy any defects that were apparent. They were media of communications and it was not long before this aspect became very important.

This leads naturally to a consideration of the state of affairs which, it was thought, was revealed by the Exhibition. To make a proper comparative survey of the processes and products on show would demand a study of its own and an encyclopaedic knowledge of the arts, crafts and technologies employed. Fortunately we are concerned, not to make such a comparison, but rather with the reactions of, and the conclusions reached by the observers of the scene. What is important is not so much the 'reality' of the situation—however that is defined— but the reactions of men to that situation.

Let us try to imagine the reflections of an informed, intelligent and reasonably impartial visitor to the Great Exhibition. He would know, before setting out on his tour, that the key factors in the astonishing development of British industry were the availability of cheap and copious power together with the skills necessary to apply that power to an enormous variety of purposes. Had not Justus von Liebig remarked that 'civilisation is the economy of power'? It was an exaggeration, of course, but not without an element of truth.

The chief agent of mechanical power was unquestionably the steam engine. Nor could there be any doubt that, from the very first successful heat engine made by Thomas Newcomen in 1712 right up to the very latest models on display in the Crystal Palace, all the major advances together with the vast majority of detailed evolutionary improvements had been due to the genius of English and Scottish engineers. Watts' condensing steam engine, his invention of the parallel motion and of expansive operation; the compound engines of Jonathan Hornblower, Arthur Woolf and William McNaught; Trevithick's high pressure engine and his steam locomotive to be followed by those of Blenkinsop, Timothy Hackworth, Robert Stephenson and many others: these were the milestones in the progress of steam power. At the same time British engineers had solved the multitudinous problems of applying steam power to the propulsion of ships and an ever increasing variety of manufacturing industries. There had been nothing like this in the whole history of civilisation.

Accompanying, and indeed making possible the development of the steam engine was the rise of the new machine-tool industry.* Again, this was mainly a British achievement. Beginning with the work of the master craftsman Joseph Bramah, a succession of empirical inventors, many of them trained by Bramah himself, had established the main forms of *industrial* machine tools [5]. Maudslay, Fox, Clement, Richard Roberts, Nasmyth and finally Joseph Whitworth had, between them, invented and developed the industrial lathe, the planing machine, the slotting machine, the radial drill, the multi-spindle drill and the steam hammer. Roberts had even succeeded in designing and making a punched-card controlled machine-tool.

Outwards from these key industries the prospect, wherever one looked, was one of assured progress: radical invention alternating with sustained evolutionary improvement. And, in spite of the earlier forebodings of men like Babbage and Davy, the British contributions to the intellectual or scientific bases of the new mechanised industry were far from negligible. Watt's friend, Joseph Black, had, in effect, founded the science of heat and notable contributions had been made to it subsequently by William Herschel, John Leslie and John Dalton. It was, of course, true that a number of important advances had been made in France during the golden age of French science; but with the exception of those by Clément, Desormes and by Sadi Carnot, these had tended to be rather more abstract than relevant to the heat engine [6]. Finally, our visitor would surely know that some very fundamental work was, even now, being done in Manchester by Mr. J. P. Joule and in Glasgow by Professors Rankine and Thomson (Kelvin). The last had, indeed, just invented an appropriate name for the new science that linked heat and mechanics—thermodynamics.

With British policies so triumphantly vindicated in all the more obvious respects what grounds could there be for doubt and for criticism? This is a very difficult question for, after all, we know little about the social psychology of world leadership. There have, in the nature of the case, been only two or three instances of one nation or people being unquestionably the wealthiest, the most powerful and technologically the most advanced in the world. Our ignorance not-withstanding it is surely reasonable to suppose that the lonely position of world leadership carries its own uncertainties and fears; and even perhaps neuroses. Always one must keep looking over one's shoulder to see who is nearest and who are catching up. As far as England in

* The generic expression 'machine tool' was first used at about the time of the Great Exhibition.

1851 was concerned the rival to be watched would be France, with Germany and America some way behind. In his more objective moments, when memories of Trafalgar and Waterloo were not uppermost, the Englishman might reflect that in Napoleonic times the French had been skilful, courageous and scientific enemies who had defeated the armies of Europe and whose ultimate overthrow had been brought about by exhaustion, the Russian winter and the largest and most efficient navy Britain had ever possessed. Moreover, by 1851 the impressive achievements of the golden age of French science could be appreciated and while it might appear that French theoretical physics had lost some of its impetus [7] this was by no means necessarily a permanent state of affairs and, in any case, French chemistry was much superior to English. A host of old saws—'pride goes before a fall', 'clogs to clogs; in three generations', etc.—representing the accumulated wisdom of mankind combined, we may suppose, with these old memories to urge the English to pay attention to Continental developments.

In the previous year, 1850, an anonymous writer, 'Philoponos', observed that there was little juvenile labour in France and hence more time for industrial instruction. The French lead in applied chemistry did not escape him; specifically he instanced Berthollet, Monge, Prieur, Vauquelin. France had indeed paid great attention to chemistry and had applied it to the arts. Some English industrialists, like John Cockerill, had even migrated to the Continent in anticipation of an English decline following the industrialisation of Europe [8].

The Edinburgh educationist, James Simpson, believed that 'we were beaten by the French whenever manufactures particularly depended on science'; a view to which some notable members of the Society of Arts subscribed. An official Belgian observer (de Cocquiel), a careful and fair-minded man, reported to his government that English industrial supremacy was due to immunity from revolution and invasion, to liberal political institutions, to great natural resources and to commercial genius, but not to industrial skill, although British manufacturers were increasingly alive to the need for industrial instruction [9]. A special commissioner, J. A. Lloyd, F.R.S., lamented the 'long neglected education of the middle classes in science', and looked enviously at France with her magnificent schools of science, both 'pure' and applied [10].

Charles Babbage was once more inspired to write about the social relations of science. Of particular interest are the opening words of

Chapter 14 of his book: 'Science in England', he remarks, 'is not a profession; its cultivators are scarcely recognised even as a class. Our language itself contains no single term by which their occupation can be expressed. We borrow a foreign word [savant] . . .'.* The list of official scientific posts he was able to enumerate is also interesting. 'A few professorships; the Royal Astronomers; the Master of Mechanics to the Queen; the Conductor of the Nautical Almanac; the Director of the Museum of Economic Geology and of the Geological Survey; officers of the same; officers of the Natural History Museum'. These offices exhausted the State patronage of science, the measure of which could be gauged from the fact that the maximum salary paid was a mere £1,300 a year (Astronomer Royal).

After noting the great differences in wealth and rank in this country, Babbage added that 'It is without doubt very desirable that all classes should contribute to the intellectual advancement of the country. But unless different advantages are proposed to different classes, it is not possible to apply any general stimulus to all'.† That science is its own reward may be true, but as an incentive theory this involves a vicious circle, 'for it can only be known to those who have already advanced in the career of discovery'. To sum up Babbage's argument; the social and economic incentives to scientific activity in England were, he believed, too small for the maintenance of that balance between science and technology which the course of progress increasingly required.

Wherever the exact truth lies in this matter, two facts are incontestable. Firstly, the Exhibition was a great success for Britain; for free-trade, for industry and for liberal common sense. The number of medals awarded to the United Kingdom by an international jury is ample proof of the first and second, the success of the Exhibition of the third. But the second fact is less creditable, for it was shown that of all the major competing European countries England and Belgium were the only ones without organised systems of technical education.

Certainly the members of the Society of Arts were not complacent for, in the following winter they organised a series of lectures on the results of the Exhibition and the lessons which could be learned from them [11]. Following the opening lecture by Whewell, speaker after speaker emphasised the need for scientific instruction, each one illustrating his point by reference to his own profession or science. But perhaps the most eloquent lecturer was Lyon Playfair, for he had

* *The Exposition of 1851* (London, 1851), p. 189.
† Ibid., p. 198.

seen, more clearly than anyone else,* that the progress of science was rapidly becoming a prerequisite for the prosperity of a manufacturing nation: 'The cultivators of abstract science, the searchers after truth for truth's own sake are . . . the "horses" of the chariot of industry'— but England neglected the 'horses'. He believed that the Exhibition showed that wherever a nation used science in its manufactures that nation was in the ascendent; and he suggested that: 'In the establishment of institutions for industrial instruction, you, at the same time, *create the wanting means for the advancement of science in this country*' for 'the progress of science and of industry in countries which have reached a certain stage of civilisation ought actually to be synonymous expressions; and hence it follows that it is essentially the policy of a nation to promote the one which forms the springs for the action of the other, I think it, therefore, no mean advantage to this nation that the establishment of industrial colleges will materially aid the progress of science by creating positions for its professors and for those who would willingly cultivate science, but are scared from it by the difficulties they have to encounter in its prosecution'.

Unfortunately they left the last word to Henry Cole who was still, at that time, attached to the ideals of dogmatic individualism. Accordingly he could say: 'The value of science depends on its practical application and that, I submit, depends on the public want of it'.

These lectures were not the only efforts made at that time; in fact the Great Exhibition ushered in a period of greatly increased activity on behalf of science and scientific education. In the year 1851, Hugo Reid, lecturing at Nottingham, had recommended the conversion of the mechanics' institutes into colleges for the people. Reid proposed a four-year course of systematic instruction in mathematics, physical and biological science and mental and moral philosophy [12]. T. E. Cliffe-Leslie, in the following year, suggested that the Society of Arts should act as a centralising agency for the institutes; rather in the same manner as London University did for its affiliated colleges. Another progressive, Canon Richson, speaking at Manchester, urged the mechanics to convert their institute into a technical college [13]; while Lyon Playfair, addressing the radical 'Peoples' College' at Sheffield, deplored the 'intense ignorance of science' among our educated classes. Playfair went on to contrast the importance of applied science with the

* He had had the (then) almost unique experience of having been an industrial chemist as well as an academic teacher.

O.S.E.—4

fact that there were only 1,190 salaried posts available for those who forwarded the sciences, philosophy and literature [*14*].

Exhortation of this kind certainly indicates alarm on the part of those who would get things done; as a way of getting them done its value is not quite so apparent. It was naïve to suppose that a large and ill-defined group, like the 'mechanics', would be capable collectively of creating an adequate system of technical education. Only an extreme individualist could believe that the mechanics could lift themselves and, at the same time, the standards of their industries, by their own shoe-strings. On the other hand it is not difficult to believe that the mechanic of left-wing persuasion would have asked himself why he should bother to acquire a technical education when the main, perhaps the whole profit from it would go to his employer.

A much more reasonable point of view was argued in 1853 in the essay—*The History and Management of Literary, Scientific and Mechanics' Institutes*—which gained the Society of Arts Prize for its author, James Hole. Central to the main argument of this essay was a sustained and eloquent plea for State aid for the institutes. Much had been done, Hole admitted, but it had been without system because 'an unreasonable jealousy of all interference of government on the part of the people, and an almost utter indifference on the part of the government itself to its own highest duties, have, in the past, prevailed'. But, 'government is a machine, liable to defects, and sometimes to breakdown. Let us diminish its defects, improve its powers, but in the name of all experience and common sense let us not cast an imperfect tool away when we have none to substitute in its place.' Following Playfair and Babbage he believed that: 'by improving the position of the scientific teacher, we should help to extend the domain of science itself. . . . It is not likely that men will devote themselves to science when the reward of years of laborious study will most probably be poverty and neglect.' A supply of lecturers and teachers would alter this, for: 'the means of living would be found for the poor student of science and the path of honourable distinction opened to him' [*15*]. Hole argued strongly that the mechanics' institutes should be made constituent colleges of the proposed industrial university (see below).

It would not have been characteristic of the Society of Arts to confine its activities to organising lectures and awarding prizes; and so, on 18 May, 1852, the Vice-President, Harry Chester, invited the unions of institutes to a conference to discuss the ways of forwarding technical education. Two months after the conference some two hundred and twenty institutes with a total of ninety thousand members became affiliated to the Society—a development which would hardly have been possible without the prior existence of the unions. It was at this point that the Rev. James Booth began to play an important part in the technical education movement.

Like Whewell, Booth was a priest of the Established Church and, at the same time, a very competent mathematician; as his original contributions to that science prove.* He differed from Whewell in that he held a more pleasing social philosophy: he was acutely aware of the 'vast reservoir of unfriended talent' among the semi-literate classes and he knew that this was both unjust and wasteful. He recognised the growing importance of scientific and technical education and he saw quite clearly that the touch-stones of then current economic theory could no more be applied to science and education than they could to the British Museum, the National Gallery or the armed forces.

Booth had joined the Society of Arts that year with the avowed object of forwarding technical education through the agency of written examinations, and the affiliation of the unions was to provide him with his great opportunity. As a result of his initiative a small committee was set up comprising himself, Bell, le Neve Foster and Twining: all stalwart friends of technical education. They circularised manufacturers, educationists and others on the question of the need for instruction in technical science. Stressing the importance of chemistry and physics in the improvement of manufactures, they urged that, in view of our educational backwardness and the enterprise shown by our foreign rivals in the matter of technical education, the mechanics' institutes should be converted into industrial schools with set syllabuses, examinations and prizes for the most successful candidates.

Most manufacturers received the proposals favourably and the final recommendation of the committee was for the establishment of a central technical college, either in London or Manchester, to teach and to examine in science [*16*]. Examination was regarded as an excellent incentive, for it was well known that, at the universities, young men would only attend lectures when they had a direct bearing on the degree examination.

These recommendations were not allowed to rest for the Society immediately proceeded to organise a comprehensive science examination system in collaboration with some four hundred leading institutes.

* He was not, however, a Cambridge man; he graduated at T.C.D.

Although the first examination was a failure, for only one candidate presented himself, a suitably modified syllabus proved much more acceptable and in 1856 there were 52 candidates. This first successful examination was held in London; in the following year the examination was held simultaneously in London and Huddersfield, and subsequently in many other centres as well. In each year the number of candidates was greatly increased and the effective area progressively extended, ranging from 'Pembroke Dock to Ipswich; from Brighton to Newcastle-on-Tyne'.*

As the chief architect of the new venture Dr. Booth was determined that it should not fail through lack of publicity. He proclaimed to the mechanics why they should learn, what they should learn and how they should learn it. Proficiency was now to be tested by rigid examination: 'The Society of Arts does not profess to teach; it examines', and examination, as we know, is an incentive. . . [17]. A little unkindly the *Athenaeum* characterised Dr. Booth's ideas as 'the knowledge-box plan'.

The Society soon recognised the need for adequate financial incentive to bring in the candidates; they therefore drew up a 'Declaration' with the purpose of giving value to their diplomas in the labour market by affording as wide a degree of public recognition as possible and by imposing a kind of moral obligation on the employers of scientific labour.† They also used their influence on behalf of their 'graduates' and, in this way, in 1856, a successful student was appointed an assistant observer at Kew observatory. But generally the jobs which came the way of students seem to have been clerkships and, in 1861, we find the Leeds Students' Union complaining that they wanted technical posts, as in H.M. Dockyards, and were not interested in the clerical appointments offered them [18].

The lists of successful candidates reveal a very high proportion of clerks and book-keepers thus confirming the truth of the assertions made about the social structure of the mechanics' institutes from 1835 onwards. They indicate, too, the comparatively slight extent to which scientific knowledge can have penetrated the great mass of the people

---

* *Athenaeum*, 19 June, 1858.

† Among those who signed the Declaration were: the Archbishop of Canterbury Sir Stafford Northcote, Sir J. F. W. Herschel, Charles Babbage, Robert Stephenson, Edward Baines, A. W. Hofmann, Dr. Jelf (Principal of King's College, London), Sir J. W. Lubbock, as well as a large number of industrialists, bankers and engineers.

—for these were the first systematic examinations established for the non-privileged classes. They were also among the first to be open to women students; Sophia Jex-Blake was a successful candidate in 1862. There is some correlation between the percentage of clerks and the numbers taking the book-keeping examinations; towards 1870 the science subjects began to be more popular and the percentage of clerks fell. When the science subjects were dropped from the examinations in 1870 the proportion of clerks increased, while the number taking book-keeping increased both relatively and absolutely. Just before that date an increasingly wide range of occupations had been represented among the examinees: mechanics, shipwrights, artisans, millwrights, teachers, governesses, engineers, weavers, warehousemen, etc., etc.

A DIGRESSION ON EXAMINATIONS

The establishment of examinations by the Society of Arts was only one part of 'a general mania for examining everybody by means of written answers to printed questions'—as the *Athenaeum* described it [19]—which swept the country at this time. It was a remarkable social phenomenon. In a few years a comprehensive system of examinations was established affecting large sections in all classes of the community above the very humblest. In chronological order they were:

(1) In 1850, the Board of Trade examinations for Masters and Mates of Merchantmen.

(2) The College of Preceptors Examinations.

(3) The Oxford Honours Schools in Mathematics, Natural Sciences, History, Law and Theology.

(4) In 1851: the Cambridge Triposes in Moral and Natural Sciences.

(5) In 1854: competitive examinations for entry to the Indian Civil Service.

(6) In 1856: the Society of Arts Examinations.

(7) In 1856: the Inns of Court Examinations. A legal university was proposed.

(8) In 1858: the Oxford and, later, Cambridge 'Local' Examinations for schools. (Inspired by the Society of Arts examinations.)

(9) In 1858: the University of London science degrees. At the same time affiliation was virtually abolished (except for medical studies) and the examinations thrown open to all comers irrespective of whether or not they had attended a college.

(10) In 1859: the Department of Science and Art Examinations.

The written examination system was derived from university practice and in social, as in academic, affairs it constituted a reform. When fully established the system meant, other things being equal, the virtual elimination of patronage in the professions and the service of the State. Booth has as good a claim as any other person to have been the pioneer of this reform. In 1847 he had published a booklet urging the creation of what was, in effect, a Civil Service Commission [20]. Ten years later he was still an enthusiast; demanding equality of opportunity and pointing out that if employers required proof of educational attainment a supply of educated people would be created [21]. But, in later years he began to have doubts and felt that the system had been pushed too far; that over-examination was a dangerous possibility.*

NEW COLLEGES AND STATE ACTION

Before describing the first action to be taken by the State in the matter of science and scientific education, we must go back a few years to 1845, when the Royal College of Chemistry was founded. Before that year the only chemical teaching, apart from the *ad hoc* courses at mechanics' institutes, was given at the London colleges where the emphasis, when not linked to medical training, was of a 'liberal' nature. The events which were to lead to the establishment of professional chemical training took place in the early 1840s when chemistry had suddenly become a relatively fashionable science.† Liebig's visit to this country (1842) had aroused great interest; not only did he enjoy Royal patronage but his work at Giessen had raised the prestige of chemistry in the estimation of English landowners. It was on the crest of this wave of public interest that a number of distinguished gentlemen, chief among whom were the Prince Consort and Sir James Clark, the Queen's physician, founded the Royal College of Chemistry. Established by public meeting at St. Martin's Place on 29 July, 1845, the College was first opened at Hanover Square.

The new College soon enjoyed a reputation quite out of proportion to its size. The first professor was A. W. Hofmann, nominated by

* Other notable advocates of examinations were Edwin Chadwick and Robert Lowe, the Liberal politician. Lowe played an important role in drafting the Government of India Bill of 1853 which, among other things, instituted competitive examinations for entry to the Indian Civil Service [22]. Lowe's influence on certain scientific matters will be discussed below.

† The Chemical Society was founded in 1841.

Liebig, who rapidly gathered together a promising group of chemists: Warren de la Rue, Perkin, Frankland, Nicholson, Odling and many others. Hofmann naturally imported German university principles, and the practice of *Lernfreiheit* was therefore adopted. Students leaving the College on completion of the course were awarded either Certificates of Attendance or Testimonials of Proficiency, the latter being the senior award, and to obtain it, it was necessary for the student to have completed some original research, worthy of publication. It was claimed that most experiments were carried out to advance knowledge and, at the same time, to increase the student's experience [23].

A parallel development was the foundation of the Government School of Mines and Science Applied to the Arts. In 1835 Sir Henry de la Beche had suggested the formation of a Museum of Economic Geology which would also be useful for instruction. With the help of Sir Robert Peel this was done, and, in 1845, the Geological Survey and the Museum together with the mining records office were moved to a new building in Jermyn Street.* The logical culmination of this was the inauguration, on 6th November, 1851, of the School of Mines. Here, too, the staff was of first-class calibre: the Principal was de la Beche, Lyon Playfair lectured on applied chemistry, Edward Forbes on natural history, A. C. Ramsay† on geology, Robert Hunt on mechanics, John Perry on metallurgy and Warington Smythe on mining; in 1854 Forbes retired and was succeeded by T. H. Huxley. Prior to the establishment of the G.S.M. there had been no mining school in this country apart from one or two unsuccessful experiments like the one projected in Cornwall by Sir Charles Lemon. For a country whose wealth was founded largely on coal, de Cocquiel had thought this shocking; and so it was, for the toll of life was extremely heavy.

The first two sessions (1851 and 1852) of the Government School of Mines were notable for the inaugural lectures given by Lyon Playfair before audiences which included some very distinguished people. In his first lecture Playfair urged the importance of ensuring a developing harmony between the advance of 'pure' science and the progress of industry; to this end the new college could make most important contributions. The relationship between science and industry had, he believed become so fundamental that if the necessary harmony was not achieved, then: 'as surely as darkness follows the setting of the

---

* The investigations of Playfair and Faraday into the state of the coalmines had clearly shown the need for more scientific methods in that industry.

† An uncle of Sir William Ramsay.

sun, so surely will England recede as a manufacturing nation, unless her industrial population become much more conversant with science than they are now' [24]. But the second lecture, given in the autumn of 1852, was more important and aroused greater interest, for earlier in that year Playfair had, at the request of Prince Albert, toured the Continent studying in great detail the methods and data of foreign technical education. He was therefore able, in this lecture, to present his findings, backed by facts and figures, and to stress their implications [25].

In the past, he said, our great and cheap natural resources had been in our favour. But such is the development of transport and communications that, in the future, differences in natural resources will count for less and less and the race will go to whoever commands the greatest scientific skill. We have, he complained, 'an overweening respect for practice and a contempt for science'. The trouble is: 'In this country we have eminent "practical" men and eminent "scientific" men but they are not united and generally walk in paths wholly distinct . . . From this absence of connection there is often a want of mutual esteem and a misapprehension of their relative importance to each other.' Here, in a remarkable fashion, Playfair had anticipated the conclusions and almost the very words of the Haldane Commissioners,* some half a century later, when they were giving the reasons for the foundation of the Imperial College.

The development of these two colleges was a very important consequence of the Great Exhibition. The idea of an industrial university was, indeed, widely approved at that time; memorials from the industrial areas urging it had been submitted to the Commissioners of the Exhibition and we have already seen what the Society of Arts thought and did. As early as August 1851 Playfair had written to de la Beche suggesting a plan for technical schools throughout the country to be united 'with a university of mines and manufactures empowered to grant degrees and diplomas' [26], and it is clear that the Prince also cherished an idea of this nature. Fortunately there was a surplus of some £186,000 from the Exhibition and this money, together with £150,000 voted by Parliament, was used to purchase the South Kensington estate. Beside the technical university it was hoped to centralise the learned societies, art galleries, museums, etc. on this estate, thereby creating a University City on a really large scale. The plan was only partially realised, of course, for among other things

* See p. 198 below.

certain of the learned societies objected to being 'planned' in this fashion.

At the same time the State itself had begun to move on a more general scale than the foundation of one technical college would indicate. In her Speech from the Throne on 11th November, 1852 [27], the Queen had promised a comprehensive scheme for art and science on a scale that befitted an enlightened nation. On 16th March, 1853, Edward Cardwell, of the Board of Trade, wrote to the Treasury suggesting the formation of a Department of Science to be allied to the already existing Department of Practical Art. The new Department was to control the Government School of Mines and the Royal College of Chemistry (which was nationalised) as well as the government geological departments and various museums and institutes in Scotland and Ireland. Approval from the Treasury was forthcoming and the new Department was constituted with Henry Cole as Art Secretary and Playfair, appointed by Cardwell, as Science Secretary. The two science colleges were gradually transferred to the Commissioners' lands in South Kensington, the operation being completed by 1872.

The Science and Art Department was not conspicuous on the national scene for several years. It created a few science schools which, with the exception of those in Aberdeen, Birmingham, Bristol and Wigan, failed before 1860 and it also paid the salaries of a few science teachers. It was not until 1859 that expansion really began when a comprehensive minute by Lord Salisbury and C. B. Adderley was passed granting aid to science teachers under certain conditions. The new system, as it developed, was one of payment-by-results. Any person who passed the Department's examinations could, if approved, set up as a science teacher, to be paid by the Department in proportion to the examination successes obtained by his pupils. This led to some of the worst kind of 'cramming' and the mental effects on many of the pupils must have been analogous to the effects of the more material cramming on the livers of Strasbourg geese. But it probably did result in a wider and faster diffusion of some science among the people than could have been achieved by any other means for the same outlay of public money [28]. In this way, the numbers examined rose from 1,300 or so in 1861 to over 34,000 in 1870, and passed the 100,000 mark in 1887.

Unfortunately the new school of science was not, for many years, a marked success. For the first ten years of its life the average number of matriculants per annum was only twelve, while the number of

o.s.b.—4*

occasional students was only fifty-four. Of mining students who, after examination, found work in mineral or metallurgical works or in the Geological Survey, the average was only four per annum (1853–71) [29]. The number of fully matriculated students remained between forty and fifty for about twenty years. This is very surprising when it is remembered that the college had few rivals, that its staff could bear comparison with that of any other university institution in the country, and that it was government backed. But perhaps we can anticipate a little and point to a clue. In the first Report of the Department of Science and Art, Playfair and Cole wrote: '. . . before scientific instruction, either for adults or for youths can be made permanently successful, it is necessary to create a taste for it by infusing it into the primary education of those classes to whom secondary instruction in the scientific principles of their trade is necessary . . . Even before (higher institutions) can be satisfactorily established, an intermediate class of secondary schools would appear to be necessary' [30]. It was deficiency, or even absence, of primary education that had inhibited the development of the mechanics' institutes; the higher scientific institution, in its turn, was to be retarded by deficiencies at the intermediate level. Secondary education, such as it was, was very defective in England at that time, and was to continue so for many years to come; it was therefore quite natural that frequent and bitter complaints were uttered by the teachers of advanced science of that day that their students came up to college badly prepared in science.

Important as the new colleges were, they were not the only new foundations created during this, the second phase of the diffusion of science in England. In November, 1853, the Birmingham and Midlands Institute of Industrial Education was opened. This was a school of industrial science providing instruction in applied chemistry, mechanics, metallurgy, mining and ventilating, geology and mineralogy. It was intermediate in standard between school and contemporary university college and was intended for the working class. Another enterprise was the Rev. F. D. Maurice's Working Men's College; but the educational aim in this case was rather more literary than scientific. In 1855 Dr. Booth founded a trade school at Wandsworth for the sons of artisans; the subjects of instruction included chemistry, physics and the steam engine. Unfortunately the school lasted for only two years. More successful was the mining college established at Wigan in 1858; after considerable initial difficulties and set-backs the college managed to survive and is active to-day. Financial difficulties were also experienced by the London Mechanics' Institute at this time; in 1858 Lyon Playfair, at the request of the government, investigated the condition of the institute which was then near collapse ('Its downfall', remarked *The Times*, 'would be an immense disgrace'). The Committee of the institute asked for government help but this was not forthcoming.

One of the most interesting endeavours of this decade was the determined attempt made in 1857 to establish a university in South Wales. The Swansea area, with its great copper smelting works, its tin plating and, above all, coal mining, presented a wide range of technological challenges. It was hardly surprising therefore that a small community of technologists and scientists, similar to the one in Cornwall, should grow up. Many of the leading families, such as the Vivians (who came from Cornwall), the Groves, the Dillwyns and the Llewelyns, had scientific interests and in 1835 the Swansea Philosophical and Literary Society was established [*31*]. In 1841 this Society inaugurated the Royal Institution of South Wales, equipped with museum, laboratory and library. It was perhaps quite natural that the next step in this area should be the establishment of a university. The proposed Western University was to have been at Gnoll, in the Vale of Neath; had it been established it would have constituted the first university college in Wales. The purpose of the institution was to teach the practical application of science to land, manufactures and commerce as well as to prepare for the liberal professions and the public services. A very high standard was aimed at, for Arthur Cayley was appointed Professor of Mathematics and other academic appointments were of comparable excellence.*

The most important as well as the most successful of the new ventures was unquestionably the foundation of Owens' College, Manchester. The progressive spirit shown by all classes in south Lancashire has been mentioned before and has been confirmed by the census report of 1851. The idea of a university institution was never far from the thoughts of Mancunians during the first half of the century. James Heywood, brother of Benjamin Heywood, had strongly urged the foundation of a college in 1836 but it was not until ten years later, when the will of John Owens, a wealthy merchant, was published that the great opportunity occurred. Owens left nearly £100,000 to

---

* I am indebted to Dr. I. Wynn Williams, of University College, Swansea, for confirmation of this last point and for making clear to me the widespread interest aroused by the proposed university.

found a college, naming as his trustees for the purpose, the Mayor, the Dean (Dr. W. Herbert, a well-known amateur botanist) and the Manchester M.P.s, among whom were James Heywood and Richard Cobden. The college, first located in Cobden's old house in Canal Street, Manchester, was formally opened on 12 March, 1851, and was affiliated to London University two months later. Among the first professors to be appointed were Edward Frankland and W. C. Williamson.

The deficiencies in the previous education of the students soon made themselves apparent as did the inevitable shortage of money. The usual appeals for government aid (November 1852 and July 1853) were followed by the usual refusals and the outlook became so bleak that, by 1858, the *Manchester Guardian* could say that the college had failed. But this was premature. Evening classes had been introduced and were successful; the London B.Sc. proved an attractive proposition in Manchester and the chemistry department flourished exceedingly under the inspired direction of Henry Enfield Roscoe.* The staff was energetic and loyal; and the people of Manchester did not abandon their college. Conspicuous among the friends of the college were the engineers Sir Joseph Whitworth, Sir William Fairbairn and C. F. Beyer [*32*]. Beyer, who was born in Germany, was one of the founders of the Institution of Mechanical Engineers and, together with Richard Peacock, established the famous locomotive works of Beyer-Peacock.

LONDON UNIVERSITY

London University was very liberal in the way in which it 'affiliated' new colleges. In 1858, when its charter came up for renewal, the University took the next obvious step and opened its examinations to all comers. Apart from the medical schools this meant the virtual disaffiliation of all the constituent colleges and the complete loss of control, never very strong, by the London colleges over the examination boards. A material reason for this step was that it was becoming impossible to enforce any uniform standard of work on the heterogeneous and widely dispersed colleges. Nevertheless this act, which was retrograde, also had its ideological aspect—it is believed that it was strongly supported on utilitarian grounds by Grote and Lowe.† After

* Appointed in 1856 when Frankland went to South Kensington.

† The *Athenaeum* favoured the change: 'Virtue or aptitude do not reside exclusively in the enrolled colleges', and, 'it is the best thing since Brougham'. (8 August, 1857 and 3 April, 1858.)

much trouble it had to be revoked at the end of the century; but from 1858 to 1903 the University of London was neither a university, nor was it of London. One is reminded of the Holy Roman Empire.

A much more interesting feature of the new charter was the institution of science degrees. These degrees reflected the growing acknowledgment of science by a large section of the intelligentsia of the time. The establishment by geological science of the antiquity of the earth as against the chronology of Archbishop Ussher and the continuing triumphs of physical science could not, in the nature of the case, do other than endow the study of science with high prestige. Also, science enjoyed the services of extremely able propagandists: Huxley, Tyndall and others. Nor should the work of Herbert Spencer be forgotten. 'What knowledge is of most worth?' asked Spencer, 'The uniform reply is—science'. While science was evidently part of the *Zeitgeist* as far as intellectuals were concerned, it would be quite wrong to infer that England was liberally supplied with scientists. and it would be equally wrong to suppose that a knowledge and understanding of science was a prerogative of much more than a relatively small section of the population.

Two memorials on behalf of science degrees were presented to London University; the first on 8 July, 1857, and the second on 12 May, 1858; both being signed by leading scientists. As a consequence of the first petition the University instituted a committee* comprising the Chancellor and Vice-Chancellor, Dr. Neill Arnott, Mr. Brande, Sir James Clark, Dr. Faraday, Mr. Grote and Mr. Walker, and authorised them to investigate the desirability of instituting science degrees. As the creation of an entirely new system of degrees is a very unusual event it is, perhaps, worthwhile to give a brief summary of the evidence of the witnesses in order to bring to light the main ideas then current as to what should constitute a scientific education [33].

John Tyndall, the first witness, advocated a degree of specialised studies not achieved even to-day. It would, he thought, be desirable to allow students to take a degree in (say) 'Heat' with some subsidiary study of other branches of physics. To Arnott's objection that you cannot advance far in one department of science without knowing something of others, Tyndall replied that any proposed course must be restricted, or specialised, by the limited learning capacity of the human mind.

In advocating highly specialised study, Tyndall was very much in the

* The first meeting was on 27 April, 1858.

minority. His nearest supporter was Bence-Jones, who advocated a two-subject degree. Frankland, Sir Charles Lyell, Warren de la Rue and Thomas Graham all supported a more liberal syllabus, as did Mr. Justice Grove, who believed that specialisation is 'decidedly wrong'. A. W. Williamson, another liberal wanted provision for the 'logic of research' in the syllabus, while J. D. Hooker, V.P.R.S. and W. A. Miller wanted the history of science included as well.

A. W. Hofmann also expressed dislike of 'one-sidedness', for knowledge of other branches is necessary for the further prosecution of one particular branch of knowledge; he therefore favoured a three-subject degree. To which Huxley replied that he did not think that Hofmann's suggestion went far enough for he would allow only collateral subjects whereas men should study non-collateral subjects 'so as to give them a general acquaintance with science'. Huxley's ideal was to inculcate a thorough knowledge of the principles of science to be followed by specialisation at M.A. level. More, he had two general criticisms to make: firstly, that one of the great evils of the country was that science was unrecognised and, secondly, that the besetting sin of scientists was that they specialised too much.

W. B. Carpenter, Registrar of the university, also favoured a liberal syllabus and, like Hooker and Miller, would have included the history of science; for 'it is most valuable for the professional scientist'. The training must be educational rather than professional, although Carpenter believed that capitalists and industrialists were referring more and more to science: 'For chemists there is a great demand'. Was it true, asked Grote, that at many of the great iron-works a scientific chemist was employed? Carpenter did not know, but he added that Frankland, while at Owens' College, 'was continually referred to by manufacturers who had no chemist of their own'.

It appears that we are, here, approaching the dividing line between the old liberal education and the coming specialised and professional one. The general tendency was to try to reconcile the two. Accordingly the first London B.Sc. required of candidates a competent knowledge of mathematics, physics, chemistry, the biological sciences and logic with ethics. Afterwards honours could be taken in any subject and after that came the D.Sc. degree which, at that time, was of approximately M.Sc. standard, and could be taken either by examination or by research. The wide range of subjects required for the B.Sc. laid the course open to the charge of encouraging 'cramming' and, to offset this evil, it was decided that the degree was to be taken in two stages;

some subjects were to be taken at the First B.Sc. examination and the remainder a year later at the Second B.Sc. examination, upon the completion of which the candidate graduated. This division of the degree eventually led to an interesting development; the First B.Sc. later became the intermediate examination, the length of time between the two examinations was gradually increased, and the Second B.Sc. became the 'final' examination. The same regulations were imposed simultaneously on the B.A. examination.*

Dr. Carpenter was a little over-optimistic. The 'great demand' for chemists was, perhaps, rather more in the nature of demands for part-time or occasional advice than the offering of salaried, full-time posts for professional chemists. Certainly Lord Granville, the University Chancellor, was more cautious. The following year we find him expressing the hope that the new degree would give a great impetus to science and scientific education. For he believed 'that science did not hold such an estimation in the public mind as to lead to the education in science which was so much to be desired'. The development of the London B.Sc. was a slow business, but the degree enjoyed great prestige in the educational world, where for many years it had no serious rival.

THE LONDON COLLEGES

Both University and King's College had opposed the disaffiliation of 1858. In the case of the latter it meant that the college drew even further away from the university and very few graduates from King's College appear in the University Calendar from 1858 onwards. The college established its own examination† which, it was hoped, would soon be recognised as the equivalent of a degree; unfortunately the public would not accept it as such. Furthermore, the college was burdened with debt and, much as the mechanics' institutes had been forced to do, it had to provide the public with what it wanted. To this end evening classes were established in the 1850s and these, for many years, formed an essential part of the college economy. Yet in spite of all these troubles, a very creditable standard was maintained: for example, in the early sixties the Professor of Natural Philosophy was

* The *Athenaeum* approved of the new degrees, commenting that the inclusion of logic and ethics was 'a step ahead of Cambridge' but, at the same time, regretted that the history of science had not been included in the syllabus [34].

† 'Associate of King's College': A.K.C.

none other than James Clerk Maxwell and when he left in 1865 his successor was W. Grylls Adams, the brother of J. C. Adams. In fact, King's College was able to establish, in the nineteenth century, a highly respectable tradition in physics and in the allied fields of engineering.

The financial state of University College was much healthier than was that of King's. Excellent scientific work was being done at that college also: in Roscoe's words 'it was at the heyday of its usefulness'. The full repeal of the Test Acts was yet to come and the college still contrived to keep some shadowy semblance of relationship with the London University.

THE OLDER UNIVERSITIES

The movement for reform of the older universities had by this time gained great momentum. There was a mounting agitation for the admission of dissenters to university degrees; an agitation which many university men honourably supported, although in 1845 Sir Charles Lyell despaired of university reform save by means of a Royal Commission. Three years later, in 1848, a memorial signed by a large number of university men and by Fellows of the Royal Society, was submitted to Lord John Russell strongly urging an enquiry.*

In the event, the government instituted the two Royal Commissions into the states of Oxford and Cambridge universities. But by this time there was a marked stepping up of the tempo of reform within the walls. We have seen what Wratislaw thought of the dominance of mathematics; two years previously, in 1848, he had pleaded for a broader syllabus suggesting that before a Tripos course be embarked upon, the students should be required to take a general course of education to include 'a popular knowledge of the history of mathematics and of the elements and history of the mixed mathematics and natural sciences' [36]. Another reformer Hugh Wyatt, also wondered whether mathematics was not being over-done: 'But does it follow that the mind's collective powers are invigorated, its general scope extended, by an exclusive devotion to such a course of discipline? . . . Is there no fear that the mind, though to a certain extent thus quickened, may at the same time become narrowed and its general character

* Among the signatories were Nassau Senior, Henslow, Baden Powell, Lyell, Charles Darwin, de Morgan, Babbage, Wheatstone, Thomas Graham, Sabine, Grove, de la Beche, W. A. Miller, Sir B. C. Brodie, P. M. Roget, Brewster, Robert Brown [35].

deteriorated?' He wanted a more extended, less narrow form of training, and to this end advocated the inclusion of the social sciences and the history of science together with modern languages [*37*]. The author of *The Next Step* . . . was of a similar mind [*38*]. He was suspicious of the concept of discipline: 'to say that discipline is the object of a place of education, is much the same thing as to say that the object of an army is to be drilled'. Like the others he wanted a general course of education to precede the taking of the Tripos.

The first major innovations were the Oxford Honours Schools in Theology, Law and History and Natural Sciences in 1850. At Cambridge the Natural Sciences and Moral Sciences Triposes were inaugurated in 1851; these were not, at first, degree examinations, but were taken after graduation. For the Natural Sciences Tripos the subjects prescribed were anatomy, physiology, chemistry, botany and geology; a curriculum as broad as that required for the London degree.

While it was generally felt, at Cambridge, that the dominance of mathematics had, perhaps, been carried too far [*39*], the institution of these new Triposes did not result in an appreciable liberalisation of the Mathematics Tripos; on the contrary, the duration of the examination was greatly increased and it was doubled in difficulty (Todhunter). Paradoxically, however, the Tripos still retained a strongly liberal flavour: the Senior Wrangler of 1868 was later to become a Judge of Appeal. The Natural Sciences Tripos was, in like manner, liberal; by no means did it produce the specialised professional scientist; for men whose chosen vocations were the Church or the law saw nothing incongruous in taking this Tripos. A practice that annoyed Dr. Booth* who, while allowing that the principle of competitive examination had shown itself eminently capable of extension to the masses, complained, with some justice, that Cambridge mathematicians went into the law or the Church. But for this 'folly', he said, their mathematical talent would serve the country.† The first generalisation was not allowed to pass unchallenged for W. Bridges Adams, a well known engineer, attacked this thesis in particular and examinations in general when he alleged that just as public bodies represent the average, so examinations would tend only to induce orthodoxy [*40*].

The mid-century was undoubtedly the great period of the Mathe-

---

* Address at Manchester Mechanics' Institution, 19 October, 1856.

† While this was not the fault of Cambridge, Booth's observation does show excellent judgment.

matics Tripos. While the technical details of the syllabus [*41*], etc., are outside our present scope, it can be asserted that no single course of instruction can, over a comparable period of time, rival the talent, even genius, that ornamented the Tripos lists between 1830 and 1870. The names of a few Wranglers are sufficient proof of this: George Green (1837), G. G. Stokes (1841), Arthur Cayley (1842), J. C. Adams (1843), William Thomson (1845), J. C. Maxwell and E. J. Routh (1854), R. B. Clifton (1859), Lord Rayleigh (1865).

The reforms initiated by Babbage, Herschel and Peacock had certainly been successful. The earlier critics, could they have forseen the events to come, would have found it hard to prove their case against mathematical education. Indeed, the middle years of the nineteenth century constituted one of the greatest periods in the history of British mathematics and theoretical physics; there was, for example, a very notable school of Irish mathematicians, led by Sir William Rowan Hamilton and the relationship between the two main schools was epitomised by the *Cambridge and Dublin Mathematical Review* (1841).

SOCIETIES AND THE STATE OF SCIENCE

The flood of reform, described above, did not leave the two major scientific organisations unmoved. At the British Association meeting at Glasgow in 1855 the Parliamentary Committee presented their report on 'whether any means could be adopted by the government or parliament that would improve the position of science or its cultivators in this country'. It was urged that all graduates should know some science (here, Oxford was felt to be an offender), and that at least the same should be expected of those who direct the affairs of the country. In order to increase the flow of university-type trained men, W. R. Grove and William Tite, M.P., demanded the establishment of State-endowed chairs in London and provincial institutions.

The committee regretted that there were so few inducements for young men to study science. Those who did, usually cast it aside at 21 or so (e.g. the Wranglers) to enter the old-established professions. Sir Philip Egerton suggested, and the Committee agreed, that it was desirable to establish a government Board of Science. The final recommendations were given under ten headings and included the further reform of the universities so as to include science in the syllabuses for all students, the comprehensive establishment of full-time institutions in all the major towns with adequate salaries and pensions for the professors thereof, the creation of a scientific civil service together with the formation of a Board of Science to control the distribution of endowments and funds.

There was, thought the committee, a feeling, widespread throughout the country, of the importance of science; but 'owing to the system which prevails in this country, of each successive government striving to outvie its predecessors in popularity by the reduction of public burdens, there is a temptation sometimes to withhold grants which may swell the total outlay of departments in which reductions are contemplated'. This, thought the committee, made the proposed Board necessary.

Two years later the President and Council of the Royal Society presented a memorial to Palmerston on the best ways to promote the study of science. Substantially their proposals were much the same as those of the British Association; including such ideas as government and local grants for science teachers, etc. They also concurred in the need for a government Board of Science; suggesting that the President and Council of the Royal Society should be recognised, at least temporarily, as such a body.

These remarkably far-sighted proposals from the British Association and the Royal Society were backed very enthusiastically by the Society of Arts; and an evident sympathiser was revealed when, in 1859, the Prince Consort could say, in the course of his Presidential Address to the British Association at Aberdeen, 'We may be justified in hoping . . . that the Legislature and the State will more and more recognise the claims of science to their attention. . . .'

Ten years later, Leone Levi, an economist, was able to show that there had, in fact, occurred a considerable diffusion of, and advance of public interest in, science since the time when Babbage wrote his *Reflections on the Decline of Science* [42] Levi estimated that 15 per 10,000 of the population contributed in 1868, either by learning or by wealth, to the advancement of science. Making allowance for duplication, he concluded that there were 45,000 persons engaged in science in this way—a total which, at first sight, is extremely imposing. But when we analyse the figure it becomes a little less impressive. Thus a large proportion of the 45,000 were members of the Royal Agricultural, the Royal Horticultural and other professional societies which were not primarily scientific in the proper sense of the word. In 1867 the membership of all the societies which dealt with mathematics and the physical sciences stood at only 3,520 with a gross annual income of £5,380. The membership for the biological societies, on the other hand, was 17,924 with an annual income of £54,614; and for the geographical and archaeological societies the corresponding figures were 7,352 and £9,601. Certainly it cannot be said that the physical sciences were well patronised or richly endowed. For the individual societies the membership figures are: Royal Botanical Society, 2,422; Royal Zoological Society, 2,923; Anthropological, 1,031; Ethnological, 219; Entomological, 208. Of the physical science societies, the Royal Astronomical had 528 members; the Meteorological, 306; while the Chemical Society was the smallest of all with a mere 192. There was no physical society. Clearly this pattern bore little relationship to the economic and social importance, actual or potential, of these sciences; and it is equally evident that the greater part of that which Levi classified as science was not only an amateur activity but, in many respects, a dilettantist one as well. The very important chemical industries were less prominent in the world of science than were the activities of amateur entomologists and orchid fanciers. But Levi was justified in claiming some progress: when Babbage wrote his first book the Chemical Society did not exist. And, in spite of his optimism, Levi found cause for dissatisfaction with the relationship between science and the state in this country.

Two years previously Grove,* too, had mentioned the obligations of the State towards science: 'To assert that the great departments of government should encourage physical science may appear a truism and yet it is but of late that it has been seriously done . . . in a time . . . short in the history of a nation, a more definite sphere of usefulness for national purposes will . . . be provided for those duly qualified men who may be content to give up the more tempting study of abstract science for that of its practical application' [43].

It is instructive, in the light of Grove's remarks, to consider an act of German policy towards scientific education in the 1860s. The Prussian government having decided that the facilities for advanced chemistry at the universities of Berlin and Bonn were inadequate, resolved to build new laboratories. These attracted attention all over Europe and with some reason, for at a time when Owens' College was still installed in Cobden's old house, the new German laboratories were built on a palatial scale. The one at Bonn for example, although ostensibly for 60 students could easily accommodate many more; it

* In the course of his Presidential address to the 1866 British Association meeting.

was equipped with every facility for the advancement of chemical science—and even possessed some that were not, such as a ballroom [*44*]. As far as the new Berlin laboratory was concerned, its main claim to distinction was, perhaps, through the person who was its first director; for in 1865 A. W. von Hofmann was called from London to take the chair of chemistry at Berlin and there he remained until his death in 1892.

Germany was not the only country that could boast a progressive scientific and educational policy. In 1862 the Federal Government of the United States passed the Land Grant Act by which colleges of 'agriculture and mechanics' were endowed by free awards of land. By 1880 there were 45 of these as well as many other colleges and universities outside the scope of the Act. But of all American foundations of this period that of the Massachusetts Institute of Technology— M.I.T.—in 1865 was perhaps the most interesting and important. Whereas the founders of the older American universities may have owed something to Oxford, Cambridge, and the Scottish universities, William Barton Rogers, the founder of M.I.T., took his inspiration from places like the Warrington Academy and from the explicit aspirations of such groups as the early members of the Manchester Literary and Philosophical Society (page 23). In other words the ideal was a utilitarian education tempered by a modern, liberal syllabus [*45*].

If we seek the reasons why America attached so much importance to education at all levels we shall find, surely, that they were much the same as those that determined German and Scottish attitudes. In Germany and Scotland educational institutions in general and universities in particular were essential for the preservation of the unity of national culture. In America education was of cardinal importance for it was the only way whereby unity could be forged and maintained by a people whose ethnic, cultural and national backgrounds were so diverse and whose number was rapidly increasing every year as new immigrants arrived.

APPLIED SCIENCE IN A NEW CONTEXT

The textile industries of Britain had been subjected to frequent and radical innovations from the very outset of the industrial revolution. They had experienced revolutionary developments in water and steam power, in the design and structures of mill buildings [*46*] and in the use of bleaching powders based on the discovery of chlorine. A Salford cotton mill was the first factory to be lit by gas light and the method of calico-printing by means of rotating cylinders represented, in Andrew Ure's opinion, the most perfect union so far achieved between chemical and mechanical science. In one respect only did the textile industries of the early nineteenth century appear to have something in common with the old, craft-based art: the dye-stuffs used were still the traditional ones. Logwoods from Central and South America, madder and indigo from Europe, Asia and Africa were the main sources of dyestuffs with woad, weld, fustic, cochineal, cudbear, lac and sumach among the other substances in common use. Using recipes based on craft knowledge and handed down from master to apprentice, dyers could produce an astonishing and very satisfying range of reasonably fast colours and shades.* Furthermore, the dyeing industry was efficient. It was based on factoring, rather like the modern tea trade, and thanks to considerable commercial skills combined with rapidly improving steam-ships, the price of natural dye-stuffs remained low throughout the nineteenth century [47].

While this was the situation and while vegetable dyestuffs remained immune from the kind of pests that afflicted cotton, tobacco and potato crops there could be little scope, comparatively speaking, for the application of chemical science to either the product or the process side of dyeing. Even so, dyers were very willing to accept new scientifically compounded dyestuffs on the rare occasions when they appeared on the market. Thus in 1822 Hartmann of Munster introduced a new bronze colour metallic dye, derived from manganese, and a year later John Mercer was using a manganese bronze dye at Oakenshaw. According to Edward Baines this dye was both fast and cheap and it was used extensively in calico printing. In 1840 chromic oxide dye was introduced in Bohemia and rapidly became universal in print works. But of all the new dyes introduced during the first half of the nineteenth century the most interesting was *murexide*. This originated in a series of researches carried out by William Prout and described in a paper read to the Royal Society. In the course of this paper Prout suggested that a substance that Wollaston had called purpuric acid might make a useful red, pink or purple dye. Liebig and Wöhler extended Prout's researches and took up his suggestion by trying to use a derivative of purpuric acid, which they called murexide, to dye

---

* The study of the surviving pattern books of the period 1790–1840 can be very revealing. The colours are excellent and the designs varied and subtle.

silk. But murexide and its related compounds such as alloxan remained of academic interest only until 1851 when Saac, in France, showed that alloxan could be used to dye silk and wool and to print calicos. In England at about the same time Edmund Potter produced a dye that he called Tyrian purple from the same source while Robert Rumney set up a factory in Manchester for the commercial production of murexide [*48*].

The reason for the delay in exploiting these new dyestuffs was that the basic raw material, uric acid, did not become available in commercial quantities until the trade in Peruvian guano began after 1840. In the meanwhile, however, organic chemists, mainly German, were studying the composition of the common natural dyestuffs and by 1850 attempts were being made to synthesize alizarine, the colouring matter of the madder root [*49*]. It took eighteen years to achieve this, but before it happened two advances were made which, while of minor importance scientifically speaking, were to mark the beginning of a new stage in the development of applied science. In 1856 William Perkin discovered the first of the aniline dyes, the so-called mauveine; three years later Emile Verguin discovered an aniline dye which he called 'magenta' in honour of one of Napoleon III's victories.

William Henry Perkin was only eighteen when he made his first important discovery. A student of Hofmann's at the Royal College of Chemistry, his ultimate objective had been the synthesis of quinine when he obtained, in the course of some experiments on aniline, a black substance from which he could extract a promising purple dye [*50*]. Realising its significance and encouraged by a favourable report from a firm of dyers, Perkin, his brother and their father, who had been in business as a boat builder, launched out as manufacturers of the new dyestuff. Technologically the moment was propitious: Mansfield, another student of Hofmann's, had just discovered a process for separating benzene from coal tar; Zinin, a student of Liebig's, had found a way to reduce nitrobenzene to aniline and Béchamp had greatly improved the process. Economically and socially, too, the times were favourable: the rise of the gas industry meant abundant supplies of coal tar and the spread of gas lighting made new, bright dyes very desirable; finally, by a fortunate coincidence, purple happened to be a fashionable colour.

Nevertheless Perkin deserves the greatest credit, not only for his original discovery but also for what Sir Robert Robinson calls 'his active, forceful pioneering' in the new manufacturing techniques of coal tar colours.* Hofmann had strongly advised Perkin against throwing up his academic career and, in any case, no one in the Perkin family had had any experience of chemical industry, much less of the new techniques that would be required to manufacture the new dyes. In spite of all this their factory at Greenford Green, near Harrow, prospered until in 1874 they sold it and Perkin retired, a wealthy man, to devote the remainder of his life to academic research. By this time a number of firms were engaged in the manufacture of aniline dyes: Read, Holliday, of Huddersfield, established as a chemical manufacturer in 1830, began to make magenta in 1860; in the same year Roberts, Dale and Co., who employed Caro and later Martius as chemists, commenced the manufacture of dyestuffs. A year later Simpson, Maule and Nicholson commenced operations and in 1864 Ivan Levinstein established his works in Manchester. In 1868 Williams, Thomas and Dower was founded and Simpson, Maule and Nicholson became Brooke, Simpson and Spiller under which name they bought Perkin's works six years later. These firms made use of the services, either as employees or consultants, of chemists of the highest ability but they were small private enterprises and no large, limited companies were formed. Manufacture, therefore, was always on a small scale [51].

The new discovery was seen to be potentially very important for here was a dyestuff that could be made from one of England's most abundant raw materials—coal. For this reason the aniline dyes occupied a place of honour at the International Exhibition held in London in 1862. As the official handbook put it: 'It is impossible to over-estimate the importance of coal-tar dyes to this country. From having the sources of raw material in unlimited quantities under our very feet, we are enabled to compete most favourably with Continental nations in this respect, and we shall soon become the great colour exporting country instead of having, as hitherto, to depend on Holland and other countries for our supply of dyestuffs' [52].

England was not, however, the only country to take an active interest in the aniline dyes. French chemists, like Delaire, Girard and, of course, Verguin, had made notable contributions to this branch of chemistry, a chair of organic chemistry had been established in Paris in 1853 and moreover France possessed a great tradition in the arts and crafts of dyeing. But it was in Germany that interest was greatest and, as events were to show, most effective. In 1862 the firm of Meister,

* Sir Robert Robinson, *Endeavour*, Vol. xv (April, 1956), p. 94.

Lucius & Brüning established their synthetic dye works at Höchst, between Frankfurt and Wiesbaden, and in 1865 the Badische Anilin und Soda Fabrik founded a works at Ludwigshafen (Mannheim). Subsequently factories were established at Elberfeld, Berlin and elsewhere. This is hardly surprising for, as we have seen, although the actual discovery was made by an Englishman, strong threads from Giessen were woven into the pattern of this new chemical industry. Many German chemists had worked, or were working in the relevant or collateral fields of chemical science. Among these men were Runge, Mitscherlich, Lauth, Caro and Peter Griess who, in 1862, discovered the first of the azo dyes which opened up enormous prospects of further synthetic dyestuffs: provided, of course, that chemists could be found to do the necessary research and that chemical engineers and other technologists were available to devise large-scale production processes. Finally, in 1868 Graebe and Liebermann announced, one day ahead of Perkin, the synthesis of alizarine from anthracene: an achievement that ultimately destroyed the French madder growing industry and incidentally had the tragically ironic consequences that the trousers worn by French infantry in 1914 owed their characteristically bright and conspicuous red colour to German synthetic alizarine.

The aniline dyes were the first of the synthetics which are now so common and they originated in scientific research organised on modern lines: they were, as we can see, the harbingers of present-day practice. It was appropriate and, at the same time, accidental that it should have begun in England; appropriate because England had the largest and most progressive textile industry in the world, a plentiful supply of the necessary raw materials and capital for the exploitation of new inventions. It was accidental because Perkin, like his contemporaries Mansfield and Thomas Anderson, worked in a German tradition; the context of his researches had been determined by the practices of Giessen, by the ideas of Liebig and Hofmann and by the contributions of many other German organic chemists. Indeed organic chemistry hardly existed in England at that time outside the German-inspired and German-run Royal College of Chemistry. Perkin's real importance lay in the fact that his work after 1856 epitomises very neatly the process of innovation based on organised scientific research; the complete realisation, perhaps, of the Baconian dream. The later history of the aniline dyestuffs industry in England was to be no less informative, if less flattering to national self-esteem.

SUMMARY

The broad outlines of the organisation of science in England first became apparent during the middle decades of the nineteenth century. As W. H. Brock has pointed out, the facilities for laboratory training in chemistry had been greatly expanded in the 1840s by the establishment of the Royal College of Chemistry, the Putney Engineering College and the Pharmaceutical Society. The following decade witnessed the inclusion of the progressive sciences in the examination syllabuses of the older universities, the foundation of the South Kensington colleges and Owen's College, the introduction of the London science degree and the beginning of State aid to science through the agency of the new Science and Art Department. The common denominator of these various enterprises was the institution of written examinations: these, as Weber noted, are associated with the 'expert', with intellectual discipline and ultimately with professionalism.

Conspicuous features of British science in the eighteenth and over much of the nineteenth centuries were the small, informal groups—reflecting, no doubt, the unorganised state of science—that came together, usually for no more than a few years, either to carry out scientific work or to achieve specific reforms. The Birmingham Lunar Society was a well-known example of an informal group that met during the second half of the eighteenth century to discuss problems in science and technology. We have already mentioned the Manchester group that formed towards the end of the eighteenth century and was interested in scientific education as well as in the pursuit of science itself: they differed from the Lunar circle in that they succeeded in establishing the 'Lit. & Phil. Soc.' as a permanent society that continues to this day.* On the other hand the Cambridge Analytical Society came into being with the specific aim of reforming English mathematics; once this had been achieved the society dispersed. Recently R. M. McLeod and J. V. Jensen have, independently, called attention to another important group, the 'X' Club, that met from 1864 onwards [13]. This group, which consisted of nine members, existed to ensure that the claims of science and scientific education were kept before the government of the day. Among the members of the 'Albemarle Street Conspiracy', as Dr. MacLeod calls the X Club, were

* The author is Joint Honorary Secretary of the Manchester Literary and Philosophical Society.

'regulars' such as Huxley, Tyndall and Lubbock together with Edward Frankland and T. Archer Hirst.* Frankland was to play an important part in the later reform of scientific education and particularly in securing the inclusion of practical science in school syllabuses.

The state of primary and secondary education was indeed a major hindrance to further scientific development. The public schools and the old endowed grammar schools were hardly touched by science; only occasionally would a headmaster include it in his syllabus. Generally it was ignored: at Eton in the early sixties there were twenty-four classics masters, eight mathematics masters and three to teach all other subjects. This was not necessarily a bad thing in itself: the German *Gymnasia* concentrated on classical education at the very time when the German universities were establishing a commanding lead in science. But it did mean that there were few posts in secondary education for would-be science teachers. There were none available in primary education, such as it was, while technical education remained moribund following the virtual collapse of the mechanics' institute movement.

The rationale of examinations is, however, independent of the popularity or otherwise of particular subjects. In 1862 the British Association met in Cambridge. There was, Edwin Chadwick discovered, 'gratifying unanimity' among the educationists assembled there that studies should be 'narrower and deeper'; and to this practice the Cambridge 'Locals' tended. Further, the masters of the special schools that had sprung up to tutor youths for the new Civil Service and Army examination were agreed that the number of subjects in examinations must be reduced so as to avoid 'cram' [54].

The unpleasantly expressive word 'cram' seems to have come into general use in, or just before, the 1840s in connection with the rapidly developing university examinations. It is not, as Todhunter remarked, a word to which a precise meaning can be attached, but it may be taken to imply the uncritical and mechanical assimilation of facts for the purpose of remembering enough to satisfy examination requirements. It was felt that when the teacher controlled the examination 'cramming' would be reduced. But whether these ideas and practices

* T. A. Hirst, Professor of Mathematics at U.C.L., edited and published in one volume Clausius' nine memoirs on thermodynamics together with the mathematical appendices (1867). The translator was Tyndall and the volume gives the clearest account in English of the establishment of thermodynamics. It is regrettably little known.

would be able to satisfy the needs of an age with new, science-based industries like aniline dyestuffs was a question that only time could resolve.

REFERENCES

[*1*] Thomas Coates, *Report on the State of Literary, Scientific and Mechanics' Institutes* (S.D.U.K., London 1841).

[*2*] T. E. Cliffe-Leslie, *An Inquiry into the Progress and Present Conditions of the Mechanics' Institutes* (Dublin Statistical Society, 1852).

[*3*] *Report of Census, 1851, Education Supplement.*

[*4*] J. B. Bury, *The Idea of Progress: an Inquiry into its Origin and Growth* (Macmillans, London 1920) p. 324.

[*5*] For the history of machine tools see the monographs by Robert S. Woodbury (M.I.T. Press, Cambridge, Mass. 1958–) and L. T. C. Rolt, *Tools for the Job* (Batsford, London 1965).

[*6*] For the relationship of Carnot's work to French science and to British technology see D. S. L. Cardwell, *From Watt to Clausius: the Rise of Thermodynamics in the Early Industrial Age* (Heinemann, London 1971) and Robert Fox, *The Caloric Theory of Gases from Montgolfier to Regnault* (Oxford University Press, 1971).

[7] J. W. Herivel, 'Aspects of French Theoretical Physics in the Nineteenth Century', *B.J.H.S.* (December, 1966) iii, p. 97.

[*8*] 'Philoponos', *The Great Exhibition of 1851* (London 1850).

[*9*] De Cocquiel, *Industrial Instruction in England. Being a Report to the Belgian Government*, trans. Peter Berlyn (London 1853) p. 5.

[*10*] J. A. Lloyd, *Proposals for Establishing Colleges for Arts and Manufactures* (Privately printed, 1851).

[*11*] Society of Arts, *Lectures on the results of the Exhibition*, (2 vols., London 1852).

[*12*] Hugo Reid, *Journal of the Society of Arts*, i, 23 September, 1853. Ibid, 9 June, 1854.

[*13*] Canon Charles Richson, *Education in Trade Schools Necessary to Promote National Education* (Manchester 1853).

[*14*] Lyon Playfair, *Science in its Relation to Labour* (London 1853).

[*15*] James Hole, *An Essay on the History and Management of Literary, Scientific and Mechanics' Institutes*, Royal Society of Arts Prize Essay (London 1853) pp. 111–12.

[*16*] Society of Arts, *Report on Industrial Instruction* (London 1853).

[*17*] Rev. James Booth, *How to Learn and What to Learn* (London 1856).

[*18*] *Journal of the Society of Arts*, ix, 12 July, 1861.

[*19*] (Augustus de Morgan) *Athenaeum*, 6 December, 1856.

[*20*] Rev. James Booth, *Examination the Province of the State: or the*

outlines of a practical system for the extension of national education (London 1847).

[21] Rev. James Booth, *J.S.A.*, v, 23 October, 1857.

[22] A. Patchett Martin, *The Life and Letters of the Rt. Honourable Robert Lowe* (2 vols., London 1893) ii, pp. 421–30.

[23] *Reports of the Royal College of Chemistry, 1845–1847* (London 1849).

[24] Lyon Playfair, 'The Study of Abstract Science Essential to the Progress of Industry', introductory lecture at the Government School of Mines, session 1851–2. Printed in *British Eloquence: Lectures and Addresses* (London 1855) i.

[25] Lyon Playfair, *Industrial Instruction on the Continent,* introductory lecture at G.S.M., session 1852–3 (London 1853).

[26] T. Wemyss Reid, *Memoirs and Correspondence of Lyon Playfair* (London 1899) pp. 134–5.

[27] *Hansard.*

[28] (A. H. D. Acland and H. L. Smith), *Technical Education in England and Wales, a Report to . . . the National Society for the Promotion of Technical and Secondary Education* (London 1889).

[29] Ibid.

[30] *First Report of the Department of Science and Art* (London 1853).

[31] R. M. Barker, *Wheatstone's Work on Submarine Cables*, M.Sc. thesis in Manchester University (1970).

[32] Joseph Thompson, *The Owens' College: its Foundation and Growth* (Manchester 1886) pp. 295–6, 314, 553–7; Appendix III, p. 633. See also B. W. Clapp, *John Owens, Manchester Merchant* (Manchester University Press, 1965).

[33] *University of London committee appointed to consider the propriety of establishing a degree or degrees in science, and the conditions on which such degree, or degrees, should be conferred* (London 1858).

[34] *Athenaeum*, 25 February, 1860.

[35] James Heywood, *Academic Reform and University Representation* (London 1860).

[36] A. H. Wratislaw, *Further Remarks on the University System of Education* (Cambridge 1848).

[37] Hugh P. Wyatt, *Thoughts on University Education* (London 1849).

[38] ——*The Next Step Respectfully Suggested to the Senate of the University of Cambridge. By one of its members* (Cambridge 1849).

[39] 'Graduate', *Strictures on Granta: or a Glimpse at the University of Cambridge* (London 1848).

[40] *J.S.A.*, vi, 1 January, 1858.

[41] W. W. Rouse Ball, *A History of the Study of Mathematics at Cambridge.*

[42] Leone Levi, 'On the Progress of Learned Societies: Illustrative of the Advance of Science in the United Kingdom during the

Last Thirty Years'. Address to the Economic Section of the 1868 meeting of the British Association at Norwich.

[43] W. R. Grove, Q.C., Presidential address to the 1866 meeting of the B.A. at Nottingham.

[44] A. W. von Hofmann, 'The Chemical Laboratories of the Universities of Bonn and Berlin', *Thirteenth Report of the Department of Science and Art* (London 1866).

[45] Dr. Julius A. Stratton, 'Liberal Education and the Usefulness of Knowledge'. An address by the President of M.I.T. at the College of William and Mary, 8 February, 1964.

[46] M. C. Egerton, *The Scientific and Technological Achievements of William Strutt, F.R.S.*, M.Sc. thesis in Manchester University (1967).

[47] C. M. Mellor, *Dyeing and Dyestuffs, 1750–1914*, M.A. thesis in Leeds University (1963).

[48] C. M. Mellor and D. S. L. Cardwell, 'Dyes and Dyeing, 1775–1860', *B.J.H.S.* (June, 1963) i, p. 265.

[49] Ibid.

[50] R. D. Welham, 'The Early History of the Synthetic Dye Industry', *Journal of the Society of Dyers and Colourists* (1963) lxxix, pp. 98, 146, 181, 229. See also E. R. Ward, 'Perkin, Mauveine and Quinine', *Chemistry and Industry* (April, 1966) p. 630.

[51] R. D. Welham, op. cit. [50].

[52] *Official Handbook to the Exhibition of 1862*, i, p. 120.

[53] R. M. MacLeod, 'The X-Club. A Social Network of Science in late Victorian England', *Notes and Records of the Royal Society* (April, 1970) xxiv, p. 181. J. Vernon Jensen, 'The X-Club: Fraternity of Victorian Scientists', *B.J.H.S.* (June, 1970) v, p. 63.

[54] *J.S.A.*, x, 24 and 31 October, 1862.

CHAPTER FIVE

The Age of Inquiries:
1868–90

It has been remarked that exhibitions, national or international, are not necessarily indices of a nation's technical and industrial progress. Indeed, to form a comprehensive judgment from an exhibition would demand near omniscience. Only perhaps in the case of specific industries can fairly objective conclusions be drawn by competent, specialised experts and even in these cases it is difficult to establish whether or not the exhibits are really representative; the choicest fruit is not necessarily put in the front of the stall. While it was true that foreign countries, aided greatly by scientific education, were rapidly making up lost time and overhauling England in certain fields, A. W. von Hofmann could afterwards say of the chemical industry displays at the Exhibition of 1862: 'The contributions of the United Kingdom and, in particular, the splendid chemical display in the Eastern Annexe prove the British not only to have maintained their pre-eminence among the chemical manufacturers of the world, but to have outdone their own admitted superiority on the corresponding occasion of 1851'.

The years 1856–65 were, however, crucial and, although it is inconceivable that the technical standards of British industry could relapse in five years, the great International Exhibition in Paris in 1867 was widely believed to have revealed a state of affairs highly discreditable to England. As a result there was alarm, near-panic and a movement was initiated that was, in all respects, more far-reaching than any so far discussed. This was the technical education movement.

The Paris Exhibition took place in the spring of 1867. Among the British jurors was Lyon Playfair who, as one would expect, lost no time in trying to assess the British position and what foreigners thought about it. On his return he sent a letter to Lord Taunton of the Schools Inquiry Commission, giving his impressions and the conclusions he had drawn from his discussions with the various experts [1]. Playfair claimed that, with few exceptions, foreigners believed that England had made little progress since 1862. He found that British engineers*

* In fact one engineer, John Scott Russell, later went so far as to say: 'it was not that we were equalled, but that we were beaten, not on some points,

and chemists lamented this and ascribed the failure—if, indeed, that was the case—to the systems of technical education developed in European countries for the masters and managers of industry. For example, J. B. Dumas, the French chemist, had told Playfair that industrial education had given a great impetus to French industries while General Morin asserted that, in his view, the best workers were Austrian; the best managers French, Prussian or Swiss. There was, Playfair concluded, an urgent necessity for a governmental inquiry into the question how foreign workers were acquiring intellectual pre-eminence and how the nations were applying this scientific skill to national industries.

The theme was taken up by Lord Granville and subsequent correspondents in the columns of *The Times* [2]. Nor was it neglected by the Schools Inquiry Commission who circularised all the British jurors* at the Paris Exhibition asking for their opinions and comment. Without exception these men agreed substantially with Playfair, although David Price, Ph.D., urged that scientific instruction was not so important for workers as for manufacturers and managers; with which Henry Cole, who was not a juror, concurred. The Schools Commission, the Associated Chambers of Commerce of several northern industrial cities and that excellent historian of technology, Samuel Smiles† all joined in the demand for governmental inquiry. In December Edward Baines was asking that the South Kensington colleges be made into a technical university, and a lively discussion was held in the rooms of the Society of Arts when E. A. Davidson delivered a lecture on 'Scientific and Industrial Education'.

In response to the widespread alarm, the Society of Arts initiated a conference on the subject and issued invitations to the mayors of large towns, presidents of chambers of commerce, presidents of learned and professional societies and the City Companies (it was felt that the City Guilds could help), university teachers, school and factory inspectors and, in short, all concerned in one way or another with technical development.

This conference [3], which was well attended,‡ opened on 23 January,

---

but by some nation or another at nearly all those points on which we had prided ourselves'.

* Tyndall, James Young, J. Scott Russell, Edward Frankland, W. Warington Smythe and A. J. Mundella.

† In a speech at Huddersfield Mechanics' Institute on 31 October, 1867.

‡ Among those present were Lords Granville, Lichfield and Russell,

1868. The resolutions down for discussion included ones on the need for improving the status of science in universities and middle and upper class schools, for bringing efficient primary and secondary education within the reach of the working classes and for founding special higher technical institutions. It was also proposed to create a standing committee to investigate further and to lobby in Parliament.

Lyon Playfair, putting the first resolution, pointed out that the universities could do little. The public school boys who went to Oxford and Cambridge—only one third of the total—were ignorant of science. While there was the excellent Royal College of Chemistry and the Royal School of Mines, *the great mining industries could not supply more than twenty men a year capable of benefiting from the courses they provided*. Earl Russell, seconding, asserted that at Oxford 'there was a very sufficient staff but hardly any soldiers' (i.e., science students). Bernhard Samuelson, a wealthy ironmaster and an enthusiast for technical education, wanted the government to endow a chair of engineering at Owens' College. Indeed, while there was general agreement on the desirability of government aid, there were even some who wanted the government to lead the people. One of these was Huxley who believed that 'you could not look to the people of this country to do anything' for their education was sadly neglected and they did not know what was good for them: a startling contrast to Herbert Spencer's views.

These men were all converts and accordingly they advocated the radical reorganisation of the whole educational machinery of the country in order to achieve greater equality and more science in education. Government aid was sought and it was urged that technical education should be borne as a charge upon the rates. To consolidate and develop these proposals they appointed a standing committee of experts.* In July, 1868, this committee presented its report [4]. The document commenced by offering a definition of technical instruction that was nothing if not comprehensive: 'general instruction in those

---

Sir C. W. Dilke, Kay-Shuttleworth, Playfair, E. A. Bowring, Henry Cole, Harry Chester, Colonel Strange, Capt. Donnelly, Grylls Adams, T. W. Goodeve, Crace Calvert, Huxley, Liveing, Fleeming Jenkin, W. J. M. Rankine, Thorold Rogers, Bartholomew Price and Michael Foster.

* The members were Richard Bentley (K.C.L.), Crace Calvert, Edwin Chadwick, Harry Chester, Robert Hunt, T. H. Huxley, Fleeming Jenkin, G. D. Liveing, Thorold Rogers and Bernard Samuelson. It was later augmented to include the Archbishop of York, Archer Hirst, Leone Levi, Augustus Voelcker, J. Scott Russell and David Price.

sciences, the principles of which are applicable to the various employments of life'.

The committee members did not favour polytechnics on the German pattern. Rather they advocated the foundation of new, liberal colleges in the style of Owens', University College and King's College together with new schools to give adequate preparatory training. Such colleges should be endowed, for experience showed that no educational institution of the highest rank is really self-supporting. The higher scientific education must be tested by approved examinations, and those who believe in it 'must prove their faith by giving practical value to the certificates obtained by students. This can only be done by the employers of labour who must, at first act on faith alone. *Hitherto no class of young Englishmen trained in the manner proposed, has existed.* In order to induce promising students to follow this methodical training they must see that the few who take that course *do* find employment more readily than those who do not. The employers of scientific labour can give an enormous impulse to scientific training by showing a real preference for young men who have passed through the course of study recommended.' As the biggest employer of all, the State must give a lead in this matter.

Nor were middle and upper class schools forgotten. Science must be included in the syllabuses and education prolonged, for preference up to the age of 18. As for artisans, little can be done at the moment because of the defectiveness of primary education. Indeed, technical schools, if created, may well fail for lack of prior discipline. They agreed with Matthew Arnold that, in this respect, it was not so much technical instruction as general intelligence that was lacking. The primary need here is general education for the artisan; a thing that may be achieved if the recommendations of the Schools Inquiry Commission are acted upon.

Specifically the government can aid by improving secondary education and by forming new science schools; by collaborating with the colleges and universities in establishing diploma examinations for scientific and technical subjects such as engineering, metallurgy, mining, naval architecture, chemistry, agriculture, etc.; by giving real value to such diplomas; by endowing and otherwise aiding educational establishments and by improving primary education. The colleges could collaborate in this by increasing the numbers of fellowships etc., and the leading firms by endowing university scholarships (a remarkably forward looking proposal) and by giving recognition to the diplomas.

One industrialist, at least, had already given a lead, for Joseph Whitworth had written to the Science and Art Department* offering to endow thirty scholarships at £100 per annum for three years. The offer was gratefully accepted and Whitworth followed up with an endowment of some sixty preliminary exhibitions at £25 per annum. On 4 May he wrote to Henry Cole suggesting the creation of a faculty of engineering with government endowed chairs.†

In the following year, in May, 1869, the Society of Arts went to the extent of petitioning Parliament on the defective state of secondary education. This, they observed, necessarily obstructed the progress of technical education. At a meeting in Manchester in the following December, Edwin Chadwick called for an end to 'rule of thumbism' and referred to the Ecole Centrale des Arts et Manufactures as the 'great Owens' College of France'.

The government acted promptly when the storm,‡ generated by Playfair's letter, broke. They instructed ambassadors and consuls abroad to report on technical education in the various countries [5], as the Schools Commission had recommended that they should, and on the 24th of March 1868 they ordered a Parliamentary Select Committee under Bernhard Samuelson to investigate the whole problem [6].

This committee declared in its report that a 'hindrance second only to the defective elementary education of the pupils is the *scarcity of science teachers* and the want of schools for training them'. The education of the smaller manufacturers and managers was usually defective, for this particular group had either risen from the ranks of the workers or had suffered middle-class education. The larger manufacturers and managers of the major enterprises had usually had better secondary schooling; sometimes with the added advantage, although this was rare, of continuation education at one of the liberal science colleges. The older universities, it was felt, did not induce those habits necessary for a successful career in industry,§ and, as only four of the public schools taught some science, it was their general impression that the

* On 18 March, 1868.
† The examinations for the exhibitions included written papers in mathematics, physics, mechanics, chemistry (including metallurgy) together with practical tests on turning, fitting, filing, pattern-making, etc. Whitworth directed that eight of these exhibitions should go to Owens', three each to Oxford, Cambridge and London universities and one each to Durham, Dublin, Edinburgh and Glasgow universities as well as to individual colleges such as U.C.L., K.C.L., etc.
‡ This is not too strong a word.   § See below, p. 247.

urgent necessity for scientific education was not realised in this country. Their recommendations were more or less in line with those of the Society of Arts committee announced six months previously: better secondary education, aid for advanced colleges of science and good elementary instruction for the working classes. As for working-class technical schools, the committee could commend only one—the Bristol Trade School, founded by Canon Moseley.*

By this time even Henry Cole had been brought to believe that efficient technical education could not be achieved without government aid. Perhaps, from his position as head of the Science and Art Department he could hardly avoid reaching this conclusion. For he told the committee that recently the War Office had allowed officers of the Royal Engineers, stationed in various localities, to supervise the Department's examinations and to inspect the science schools. The reason for this innovation was that *'you could hardly find a numerous corps of scientific inspectors, at present, except in that particular body'.*† In other words there were virtually no professional scientists in England at that time. This is quite apparent from the very small number of Royal Engineer officers required: in 1879, thirty-nine officers were employed in school inspection and fifty-eight, ranging from a Major-General to Lieutenants, in supervising the examinations. The practice of employing such officers as school inspectors continued until the nineties when, at last, it became possible to dispense with their services.

Captain Donnelly, R.E., of the Department, believed that the slow growth of the School of Mines was due to defective primary education. Playfair, now Professor of Chemistry at Edinburgh, maintained that before extended scientific education, so very desirable, was possible, the educational system of the country would have to be reorganised. Secondary education should be made available to the working class and with it an adequate scholarship ladder to the liberal science colleges. The measure of inadequacy of the educational system was given by Trenham Reeks when he showed that the number of matriculants at the School of Mines in the previous year was only eighteen. Generally this small school was not full; according to Reeks, this was because the public was indifferent to science. On the other

* See above, p. 47.
† Or, as a report of the Board of Education put it, some fifty years later, the officers of the Royal Engineers 'in the early days of the Department were one of the few bodies of men in the country with an organised scientific training' [7].

hand Fleeming Jenkin reminded the committee that Zürich Poly-technic enjoyed a State subsidy of £10,000 a year and possessed laboratories superior to any in England. We could not, added Jenkin, follow suit for the education of our young men was too backward to permit them to profit from advanced courses; therefore the prime need was for improved secondary education. The universities could well supply the teachers that the new schools would require. Until that happened scientific instruction, Jenkin thought, could not be improved.

Another witness who took a liberal standpoint was Roscoe, the able and energetic Professor of Chemistry at Owens' College. Roscoe remarked: 'I do not think we ought to give scientific education to any particular class of people as a class . . .' It should be open and available for all. The primary function of the universities was, he thought, the training of teachers: 'I think at the present time that it is the great work which we, in the science department of the college (Owens') have to do; and I think that that alone would be a return for the endowment of such colleges'. For the present he feared that not only were the German universities materially far better equipped than ours, but they were also to a much greater extent permeated with 'a love of science and knowledge for its own sake'. Roscoe was perhaps over-modest; his own laboratories were rapidly gaining in prestige and he already had over a hundred students at work in them.

Edward Frankland, of the Royal College of Chemistry, informed the committee that many chemistry students had to wait two and some-times three years before they could find suitable employment. Germany and Switzerland had relatively many more teachers than England had: at Zürich Polytechnic there were sixty professors and lecturers, at Karlsruhe forty-seven, at Dresden twenty-three, at Hannover twenty-four and at Vienna fifty-seven, compared with twelve at the South Kensington college and seventeen at Owens' (all departments). Frankland was, of course, not alone in admiring the Zürich Polytechnic. A writer in the journal of the Society of Arts, a few years later called attention to '. . . the excellent organisation and general thoroughness of this great school. This country can boast of no analogous institution at present . . . The effort most heroically made by King's College to establish a polytechnic school in London has, in spite of the great energy exhibited, proved only a partial success' [8].

Dr. W. B. Carpenter provided the information that the London science degrees had not proved a great success; perhaps the requirement of Greek at matriculation discouraged such students as those who attended the South Kensington colleges from attempting the course. But the D.Sc. degree which, as Carpenter put it, '. . . would be sought rather by those who desired to become teachers' was, understandably, even more of a disappointment. One witness, J. F. Iselin, of the Department of Science and Art, estimated that few 'science' teachers had, as yet, had a special training; and only five or six had achieved a London degree. Conspicuous among the latter was Mr. Bithell, D.Sc., who was in business in the City and taught part-time in a Hackney school. More appropriately, Dr. W. M. Watts was science master at the Manchester Grammar School.

The industrial and commercial witnesses were also united in demanding better education and special schools devoted to science; they, too, generally concurred in the need for State aid. On this there was almost complete unanimity and even those like John Platt, who felt that Playfair had exaggerated the dangers, saw that it would be fatal to stand still and that rapid developments would be necessary if Britain was to hold its place in a world that was becoming highly competitive. A. J. Mundella went so far as to echo Thomas Henry's complaint of nearly a hundred years before by remarking that 'I do not think we have a single dyer in our immediate neighbourhood who is a good chemist'. James Kitson, the Leeds ironmaster, added, for his part, 'I do not know a single manager of ironworks in Yorkshire who understands the simple elements of chemistry'.

The government had, just prior to ordering the committee, considered the formation of a central university of science, using the Royal School of Mines, the Royal College of Chemistry and the Royal School of Naval Architecture (now the Greenwich Naval College) as the main component colleges. In addition, according to Lord Robert Montague, the government had been prepared to aid university science courses and also to establish provincial colleges on the lines of the Royal College of Science in Dublin [9] if local interests would also contribute. Chairs were to have been endowed and scholarships awarded to a total value of £25,000 a year. The proposed university course was to have been of three years duration; the first and second years devoted to 'pure' science and the third to industrial technology; specifically, the School of Mines was to have been the third year mining college. The Duke of Marlborough, the Lord President of the Council, agreed with the plan and a minute was prepared but it was not passed as money was said to have been short.

The idea of a technical university was not, however, allowed to rest.

A movement to promote such an institution was launched in July, 1870, and, in December of that year, a general committee was formed with the Lord Mayor of London presiding. In the following June meetings were held to support the claims for the university, with Lord Shaftesbury playing a prominent part, and a few months later another conference was held at which Antonio Brady presided. What followed this, the second sustained attempt to found a university of technology, must, however, be deferred to a later page.

In the meantime the initiator of the debate, Lyon Playfair, was continuing his efforts to proselytise. In 1870 he was lecturing in Edinburgh on technical education, pointing out that Britain was practically last in the European educational field, praising the Zürich Polytechnic and laying down that the gap between science and practice must be bridged by men having technical knowledge and special aptitudes. As for working-class technical education, he completely rejected the idea that sufficient knowledge to enable him to do his job was good enough for the artisan [*10*]. Later that year he delivered an address* at the Birmingham and Midlands Institute on his election as President, in succession to Charles Dickens. On this occasion, too, he called attention to the gap between science and practice and the need for filling it. It was in fact the need for true applied science that Playfair was so strongly urging.

THE DEVONSHIRE COMMISSION

Colonel Alexander Strange was one of that group of officers of the Royal Engineers and the Royal Artillery who, over the course of the nineteenth century, rendered notable services, direct and indirect, to the advancement of science. Born in 1818, the fourth son of Sir Thomas Strange, he went from Harrow into the Indian Army where he was engaged on survey work, particularly in connection with the great Indian Trigonometrical Survey. Of marked scientific ability, he found time to contribute papers to the Royal Society, of which he became a Fellow, the Royal Astronomical Society, the British Association and the Meteorological Society. On his return to England in 1861 he was elected to the Council of the Royal Society, to the Council of the Society of Arts and he served as a juror at the exhibitions of 1862 and 1867.

It is clear from his public speeches and writings that Strange was gifted with the efficiency of the good soldier, the imagination of the

* 'Inosculation of the Arts and Sciences.'

good scientist and the intelligence of both. It was natural therefore that when he discovered that there were, in England, virtually no public facilities for scientific research he should try to remedy the deficiency. Circumstances were favourable for, as we have seen, technical education was attracting great attention at the time. But Strange was not following the lead of others; rather, in fact, the contrary. He was not, as were many of the supporters of technical education, an industrialist, an economist or a social thinker; he was to a much greater extent a natural scientist with revolutionary ideas about the role of science in society and the correlative duties of society towards science.

He took the first step at the British Association meeting at Norwich in 1868. Strange has previously discussed his ideas with R. J. Mann who had advised him to present them in the form of a paper. The resulting essay [11], read to the Mathematics and Physics Section, propounded the thesis, startling in an age when many still hankered after unrestricted individualism, that the State alone could adequately support the advance of science. Strange advocated the foundation of a chain of research institutions; for private scientific enterprise, great as it was in England, could not meet the need. Moreover, it was his opinion that 'the tendency of progressive civilisation is to supersede individual efforts'. Whatever heresies the paper contained it must have made a great impression on the British Association for a committee* was at once appointed to investigate further and specifically to answer two questions: did there exist in the United Kingdom sufficient provision for physical research and, if not, what additional resources were needed?

In the following year this committee, now augmented to include Alfred Tennyson, F.R.S., Lyon Playfair and J. Norman Lockyer, presented their report [12]. They had found that men of science did not believe that there were adequate provisions for the vigorous prosecution of physical research. But they also remarked that any scheme for science should be based on a full knowledge of the facts and these they did not possess; nor did they see how they could obtain them. Therefore they recommended that the full influence of the Association should be brought to bear to secure a Royal Commission to examine:

* The members were: Strange, Kelvin, Tyndall, Frankland, Stenhouse, Mann, Stokes, Huggins, Glaisher, A. W. Williamson, F. Jenkin, T. A. Hirst, Huxley and Balfour Stewart. (Kelvin was, at that time, Sir William Thomson; it is more convenient however, to refer to him simply as 'Kelvin'.)

(1) The character and value of existing institutions and facilities for scientific research, and the amount of time and money devoted thereto.

(2) What changes were desirable in the means and facilities that were available.

(3) In what manner these changes could best be accomplished.

These proposals were submitted to the Council; but in the meantime another question had been raised by A. W. Williamson and W. H. Miller. This was: had the government been impartial in its aid to higher scientific education? Had their actions been such as to utilise and develop the country's resources for the free development of scientific education? So now there were two separate questions before the Council; one relating to research, raised by Strange, and one to education, due to Williamson and Miller. The Council thereupon submitted both questions to a sub-committee for further consideration. Little time was lost, however; the final report of this sub-committee was accepted and a formal request was made that the Lord President receive the Council as a deputation. This was agreed and on 4 February, 1870, they were received by Earl de Grey.

Two months later a meeting was held at the Society of Arts when Strange read a paper on the 'Relation of the State to Science' [13]. He pointed out that whereas various sums of money were disbursed by the State on such scientific institutions as the British Museum, the Botanical Gardens, the Science and Art Department, and the Surveys, as well as for specific projects such as investigations into armour-plating and explosives, it was all done without system and method. Some branches were liberally, others quite inadequately subsidised. Surely there was need for efficient organisation? Strange was addressing a very sympathetic audience, so that when Mann put an appropriate resolution* it was passed unanimously.

In the following year Strange put forward his ideas in some detail in a paper which he read to the Royal United Services Institution [14]. There were, he asserted, a number of questions of national importance which, in the nature of their cases, demanded scientific treatment. Specifically these were defence, telegraphy, meteorology, astronomy, ventilation, sewerage, public hygiene and the surveys. All these were,

* That '. . . this conference desires emphatically to affirm the conclusion of the British Association that a Royal Commission to inquire into the relations of the State to science is very desirable and to recommend that the scope of the inquiry be made as wide as possible'.

O.S.E.—5*

he asserted, immensely important and for their proper administration a science council would be necessary. The function of this council should be advisory, but it should also administer research grants and, when required, carry out experimental work. The members of the council should include the presidents of the learned societies together with military and naval representatives and elected scientists. Over all there should be a Minister of Science.

Strange was again active at the British Association meeting at Edinburgh that year. He reiterated his views in a paper entitled 'On Government Action in Science' [*15*] and this time Kelvin, in his Presidential Address, referred to the matter. The British government, said Kelvin, fatally neglects the advancement of research. In this respect it compares most unfavourably with, for example, Germany where well-equipped laboratories are available for scientists.

Before we go on to consider the Royal Commission that resulted from this movement it is appropriate to pay tribute to Strange's intelligent appraisal of the situation. He saw very clearly what relatively very few indeed saw at that time: the possibilities of and the necessity for a developed system of applied science. Perhaps only Playfair had seen this before, and as clearly as Strange did. Moreover having envisaged the possibilities, Strange bent all his energies to realising his objectives and it is a testimony to his judgment that practically everything he called for has, to-day, been realised. It is astonishing what he achieved in so few years and it is a sad commentary on the state of science in England at that time that it was an Indian Army officer and not an academic, a civil servant or an industrialist who saw so clearly what the defects were and what the remedies should be. Unfortunately Strange did not live long enough to see his ideas adopted; he died prematurely at the age of 58.

The Royal Commission of 1872, the famous 'Devonshire' Commission,* which was instituted in response to Colonel Strange's movement, was in many respects the most satisfactory of all the State inquiries into science during the nineteenth century and before 1914 [*16*]. Keeping strictly to the terms of reference and limiting themselves to a field of manageable proportions the Commissioners were able to throw con-

---

* The Chairman was the Duke of Devonshire (who, at Cambridge had been Second Wrangler and Smith's Prizeman) and the members were Huxley, G. G. Stokes, H. J. S. Smith, J. Kay-Shuttleworth, Lord Lansdowne, Samuelson, William Sharpey, W. A. Miller and Sir John Lubbock. J. Norman Lockyer was Secretary.

siderable light on contemporary science and at the same time, to make suggestions which, had they been fully implemented, would have gone far to meet the scientific needs of the country.

Considering the evidence from Oxford and Cambridge together, we learn that while many undergraduates took science as part of a liberal education, the Honours Schools at both universities had proved disappointing. At Oxford, for example, the Honours School in Modern History had grown four times as rapidly as that in Natural Sciences.* The Professor of Anatomy at Cambridge, G. M. Humphry, was disappointed with the results of the Natural Sciences Tripos which had averaged about twelve candidates a year, many of them being medical students. There were, said Humphry, not enough prospects for the scientist; a view endorsed by the Rev. William Cookson, of Peterhouse, and the Rev. Henry Latham, of Trinity Hall. A remedy for this was suggested by Jowett who was inclined to think that the only way to encourage science would be to 'attach' it to some profession; he specifically instanced medicine and engineering but he was unable to suggest how the attachment could be effected. Professionalism normally implies specialisation but no one giving evidence before the Commission advocated a specialised education in science; on the contrary witness after witness specifically repudiated such a thing,† while Huxley commented that to award a degree for proficiency in one subject would be to make a technical college of the university.

The evidence of the academics shows that although science was slowly becoming professional this was almost entirely due to the demand for teachers that was gradually making itself felt.‡ Indeed, it seems that the universities were badly served at both ends for the schools still did not provide proper scientific training and, consequently, there were far too few vocational opportunities for graduates. Latham's solution of the second difficulty was that the government should establish State technical colleges as had been done in Germany —where, he added, science was a profession that offered a very adequate livelihood. It will be recalled that this suggestion was very similar to the one made by Playfair and by Hole some twenty one years earlier.

* Evidence of Benjamin Jowett.
† Among them were Kelvin, Couch Adams, R. B. Clifton, Bartholomew Price, Sir Benjamin Brodie, Jowett and G. M. Humphry.
‡ According to the evidence of Clifton, Price, Rolleston, Challis, Cookson and Latham.

An institution providing instruction in applied, professional science was, of course, the Royal College of Chemistry. But as we have seen the number of matriculants every year was small. In fact the total of students was fairly constant between 1845–6, when it was 49, and 1869–70, when it was 41. An increase in the length of the course following nationalisation in 1853 had tended to reduce the annual entry. But in spite of this the college had done excellent work; about one hundred and forty papers had been published in scientific journals and the names of students were among the more distinguished of the time.* Edward Frankland produced details showing the occupations subsequently followed by former students:

Teachers	Pharmacists	Iron and Mines	Brewing	Govt.	India	Chemical Industry
38	38	25	27	18	29	106

Some brewers, said Frankland, even employed *two* chemists!

In Frankland's opinion, however, England was very much behind France and Germany. He supported this view by showing that in 1886 some 777 papers on chemistry were published in Germany by 445 authors (1.75 papers per author); 245 papers in France by 170 authors (1.44 papers per author) and in Britain 127 papers by 97 authors (1.31 papers per author)—and many of the last were written by Germans resident in Britain.† Not only were there more scientists in France and Germany but they were more productive: for British backwardness Frankland blamed the failure of the universities to recognise original research and the complete lack of State laboratories and State subsidies.

Although there was, therefore, some small progress being made in chemistry and a slight demand for technical chemists, there was virtually no demand for physicists, apart from some very limited opportunities in the telegraph service‡ and in the teaching profession.

The only State support for 'middle-class' science was the grant of £1,700 per annum to the examiners of London University and the more handsome allowance of £7,000 paid to the Scottish universities.

* For example: de la Rue, Nicholson, Crookes, Graham, Perkin, Frankland, Lockyer, Odling, Abel, besides the Director, Hofmann.

† e.g. Griess and Hofmann.

‡ Oliver Heaviside, a nephew of Wheatstone, began his career in the telegraph service but deafness soon forced him to resign. Heaviside (1850–1925) was a most able physicist and may be fairly described as the last amateur of science.

In spite of the absence of official support the number of science students at University College was almost equal to the totals for Oxford and Cambridge combined; while Owens' College could count more science students than either university. Also revealing was the distribution of Oxford and Cambridge Fellowships: at the former University there were nine natural science Fellowships out of a total of 165; at the latter, three out of 105.

As the main witness, Strange was given a full opportunity to present his views. Explicitly he proposed the establishment of State science laboratories to deal with astrophysics, geophysics, physics, metallurgy, chemistry and physiology. He urged the extension of the Natural History Collection, the establishment of a science museum and State endowment of the universities. All this was necessary, he argued, for we cannot tell what branch of science, now economically unproductive, may not eventually lead to untold wealth. These proposals were generally received with favour by the scientists.* Kelvin, in particular, approved very strongly, suggesting that the capsizing of H.M.S. *Captain*, the first turret ship in the British navy, during a storm in the Bay of Biscay, might not have occurred if the Admiralty had been able to refer the design of this unorthodox ship to an appropriate research laboratory.

Let us try to summarise briefly the findings of the Devonshire Commission. They took the view that education in science at the universities should not be specialised; they believed that some literary culture was desirable in a scientific education, and that some science should be taught to classicists (Third Report). They demanded the radical reform of secondary education (Sixth Report) and they dealt exhaustively with the question of the State endowment of science. On this last point '. . . the whole scientific world of England contributed its quota' and their Eighth Report was substantially a vindication of Strange's proposals: they advocated State laboratories, increased research grants for private scientists and they recommended a Ministry of Science and Education with a Council of Science to assist it.

The immediate consequences of the Devonshire Commission are, however, less easy to assess. Perhaps the most important achievement was the provision, by the government, of a grant of £4,000 a year, to

* Joule, Kelvin, Frankland, Balfour Stewart, Spottiswoode, Hooker, Burdon Saunderson, Williamson and de la Rue. Kelvin's views on the sinking of H.M.S. *Captain* were not only those of a brilliant physicist: he was a very experienced yachtsman.

be administered by the Royal Society, specifically for endowment of research. This scheme was initiated at the end of 1876 for an experimental period of five years [17]. Insignificant as the grant seems by modern standards, it represented a crucial step forward in the involvement of the State in scientific research and it indicated an official confirmation of Strange's arguments. At the same time steps were taken to improve the State-owned South Kensington colleges.

Henry Cole, as we have seen, had already abandoned the doctrinaire individualism he had professed in 1851. By the time of the Devonshire Commission he had gone so far as to write to the *Economist** to suggest a fixed government policy of investing national surpluses in science and art: 'Mr. Disraeli, the Marquis of Salisbury, Lord Derby and Sir Stafford Northcote have expressed their conviction that the progress of industry depends on the cultivation of science and art and I hope that they will act boldly now that they have the power to do so.'

And yet, in spite of the Devonshire Commission, the climate of opinion was distinctly chilly for science. The politicians, the civil servants, and the public did not rise to the occasion. Generally, *laissez-faire* and self-help continued to be the rule in science. So, when the British Association approached the government with the request that the State take over the tidal observations they had been carrying out since 1867, they met with a blunt refusal (1872). A penetrating insight into the mentality of those who opposed the endowment of science was provided by the experience of a deputation from the Scottish Meteorological Society, which included Lyon Playfair, when they waited on Robert Lowe, the Chancellor of the Exchequer, in 1869 [18]. The deputation was to request a grant to enable the Society to continue with its observations. Lowe, ever the implaccable utilitarian, read them a most severe lecture on self-help: 'I am in principle opposed to all the grants and it is my intention not to entertain any applications of this nature. We are called upon for economy . . . I hold it as our duty not to spend public money to do that which people can do for themselves.'

The grant asked for was £300 a year.

THE CITY AND GUILDS MOVEMENT

The Society of Arts had withdrawn science subjects from their examination papers in 1870 when it became clear that these were

* 21 November, 1874.

overlapping the Department of Science and Art examinations. The Society could not hope to compete with a department of State, nor was it desirable that they should attempt to. The Department, running the payment-by-results system was disbursing £40,000 to £50,000 a year on examinations while the Society could afford only £600 to £700. But on the withdrawal of the science papers the examinations lost much of their purpose and so, in 1871, it was decided to terminate them altogether in the following year. At this point Donnelly suggested that the Society could continue its useful work by instituting examinations in technology, with papers designed specifically to supplement the science papers set by the Department. It was suggested that in this work the City Livery Companies could well collaborate [*19*].

There was a general feeling throughout most of the nineteenth century that trades or technologies were not proper subjects of instruction; only the relevant sciences should be taught. In later years Sir Philip Magnus ascribed this, in so far as it was official policy for the Department, to the belief that to teach trades would violate the principle of free-trade by putting, in effect, a State bonus on the industries in question [*20*]. While this is undoubtedly true it is not the only explanation for the policy of confining such vocational education to 'pure' science; there was also the firm, and not unreasonable belief, going back to the origins of the mechanics' institutes movement, that trades and skills are best learned—indeed, can only be learned—in the workshops. There was also the fear of some employers, who felt that they had secrets to lose, that technological education would create a class of industrial spies. However, this is a case in which it is quite impossible to make fine distinctions. At what point a technology ceases to be technological and becomes scientific is never very easy to determine. One cannot, therefore, accuse the Department of inconsistency when we find that, at a later date, they had so far departed from the strict syllabus of 1859* as to include such topics as machine construction, building, agriculture, mining and navigation.

Let us, however, revert to the activities of the Society of Arts. Donnelly's suggestion was approved and, a month or two later, a small group† set to work to study the details [*21*]. In the following July there was a conference [*22*] attended by the Lord Chancellor and the leaders of the movement. Once again Donnelly's scheme was approved

* Mathematics, physics, chemistry, geology, natural history.

† W. H. Perkin, Chadwick, de la Rue, Douglas Galton, Whitworth and others.

and so were proposals that the City Livery Companies should partici-
pate. From this time onwards it was increasingly felt that the very
wealthy Guilds should justify their existence. It was recalled that one
of the original purposes of these old foundations had been the instruction of the apprentice in his craft; it was known that the idle funds of
the Guilds were more than sufficient to enable great improvements to
be made in technical education. The obligation was clear and certainly
the Guilds made no attempt to evade it; for, on 21 July, 1873, the
Prince of Wales conferred with their representatives at Marlborough
House on the best ways in which the City Companies could aid
technical education.

In 1873 Donnelly advised Sir Sidney Waterlow, the Lord Mayor,
that in view of the vast network of classes run by the Department, the
best help that the Companies could give would be to endow scholarships and bursaries at provincial institutions, like the new liberal science
colleges that were being founded at the time, and to assist in the
establishment of chairs and laboratories.

The recommendations were carried out. But, at the same time, the
idea of a technical university was kept in mind. On 28 October, 1873,
a conference of City Companies agreed to a proposal to build a teaching
institute in the City and, for this purpose a sum of £10,000 was
allocated. For a few years the matter rested and then, in 1877, it was
officially announced that a City and Guilds industrial university was
contemplated. The building was to be between the Temple and
Blackfriars and the university was to have affiliated provincial colleges
united by a system of technological examinations similar to the
industrial university proposed by Booth, Reid, Playfair and others
some twenty-five years earlier. On 14 March, Donnelly wrote a comprehensive letter to Waterlow, proposing an escalator system of
scholarships from the provincial institutions to the university. He
advised that science, 'pure' and applied, comprise the main subjects,
although it would later be appropriate to include the social sciences
and modern languages. He insisted on a good staff, sound buildings
and suitable laboratories and he maintained that practical, laboratory
teaching was essential.

In the summer of 1874 came the detailed report of the working party
which the Guilds had instituted under the Chairmanship of Lord
Selbourne. It was recommended that some £20,000 be spent as from
the following January, one half being kept for the new university and
the other half being distributed as scholarship money and as aid to

existing London and provincial colleges. At the same time a constitution was drawn up for the proposed 'City and Guilds of London Institute'.

The new institute—the 'City and Guilds'—was formally inaugurated at a meeting at Mercer's Hall on 11 November, 1878; the Prince of Wales was designated President and the three Vice-Presidents were Lord Selbourne, Sir Frederick Bramwell and Sir Sidney Waterlow. In 1879 the institute took over the technological examinations of the Society of Arts and introduced a system of payment-by-results. This had an immediate and startling effect on the numbers of candidates coming forward: in 1879, the year of transfer, there were only two hundred and two candidates, by 1883 there were well over two thousand and by 1888, over six thousand.

Unfortunately the scheme for the new buildings was not going forward as fast as many had hoped. It had been known that the Commissioners of the 1851 Exhibition funds had been considering a plan of their own for a technical university and were intending to spent £100,000 on a building in South Kensington, always provided that the Treasury would undertake to maintain the college. This the government, with characteristic obtuseness, would not do, and it became known, unofficially, that the scheme had fallen through. At this point the Guilds stepped in with an offer to build and maintain the college if the Commissioners would provide the land. This offer, unfortunately, caused more difficulties, now within the Guilds themselves for dissidents pointed out that by no stretch of the imagination could South Kensington be considered a part of the City and, after all, a City institution should, by definition, be located within the City boundaries.

On 5 December, 1879, T. H. Huxley took the chair for a lecture given by the brilliant teacher and writer Silvanus P. Thompson at the Society of Arts. The subject was 'Apprenticeship, Scientific and Unscientific'. Thompson expressed some very emphatic views. There was only one good trade school in the country; that founded by Canon Moseley. There were no great technical colleges to compare with those of France and Germany and there were no scientific managers or employers. And, he continued, '. . . there is no question whatever but that the persistent neglect of technical education in England will sooner or later ruin her in the markets of the world. . . . The skilled industries of Great Britain with their irregular bands of workers trained anyhow, nohow (sic), armed with fantastic scraps of empirical knowledge . . . are doomed . . .'

At the end of the lecture Huxley took the opportunity to utter a melodramatic threat to the Guilds. A great deal had been said about a Minister of Education and about the role of the State, but he would say that, as far as London was concerned, it would be an utter scandal and a robbery if one shilling was taken out of the general revenue to pay for technical education. There were, in the City of London, at the present moment, the possessors of enormous wealth, who were the inheritors of the property and traditions of the old Guilds of London— organisations originally intended for this very purpose—and if the people of England did not insist on this wealth being applied to its proper purpose then they deserved to be taxed down to their shoes!

Thompson's lecture and Huxley's adroitly timed threat received wide publicity. On 11 December there was a leading article in *The Times* and three days later a reply from Huxley; others joined in and the grievance was extremely well ventilated. While it is impossible to assess accurately the part Huxley's threats played in determining the course of events, it is at least notable that firm decisions were taken with gratifying speed almost immediately afterwards. At the second annual meeting of the governors of the City and Guilds Institute it was agreed that the Commissioners would provide the land while the Guilds would put up £50,000 to build and £50,000 a year to maintain the proposed institution.

In the meantime there had been running for some three years, at Cowper Street in the City, a series of lecture classes in applied science. H. E. Armstrong was responsible for applied chemistry and W. E. Ayrton for applied physics, or electro-technology. These very modest courses were the beginnings of the City and Guilds *College*; but the ultimate intention was to build a junior technical college in the City, thereby satisfying the parochial elements, and an advanced one in South Kensington. The first would cater for artisans and also act as a feeder for the South Kensington college which was to be an applied science complement to the government colleges and thus to be of university status.

On 10 May, 1881, the foundation stone of the City college at Finsbury was laid by Prince Leopold in the presence of a distinguished company. In the course of his address the Prince remarked: 'My Lord Mayor . . . I have now had the pleasure of laying the foundation stone of the first technical college ever erected in London . . .' On 18 July of the same year, the first stone of the South Kensington college was laid by the Prince of Wales.

In May 1880 Philip Magnus was appointed Secretary and Director of the City and Guilds Institute. Magnus later described some of the difficulties that the new venture had to meet and overcome [*23*]. At that time there was, he said, no country so backward in technical education as the United Kingdom. In spite of all the efforts made during the preceding half century this form of education was quite unsystematised. Indeed, the form that technical education should take had never been thought out properly as it had been in Germany and in France. This was the task to which Magnus and his colleagues addressed themselves: with the advantage over all their predecessors that there was at last adequate financial backing.

Huxley gave a definition of the aims of the movement: it was to provide theoretical and practical instruction for artisans and others engaged in industry; an adequate supply of teachers of technology together with adequate schools in industrial areas and, lastly, sufficient scholarships leading ultimately to jobs as teachers or as *original researchers in applied science*. To many, although by no means to all, technical education had meant a smattering of simple science for the working man: this, of course, was the view that had dominated the mechanics' institutes and the first examinations of the Society of Arts. In later years it became fashionable to assert that this showed that very few had properly understood the importance of scientific training for managers and employers. But this attitude is less than just to the long succession of men from Cooke-Taylor's time onwards, who saw the need very clearly; it is also unfair to such institutions as the Royal School of Mines, the Royal College of Chemistry and Owens' College.

Magnus and his colleagues were familiar with German practice and accordingly they set to work methodically to establish the new venture on as sound a basis as possible. Studies were made systematic and entry, even to Finsbury, was conditional on passing an entrance examination. *Lernfreiheit*, however, was not practised. Armstrong, Ayrton and John Perry worked hard at Finsbury to create the new technological education in detail. In Magnus' words: 'In the small laboratories of the Finsbury Technical College the beginnings were laid of that reform in science teaching which the pioneers of the technical education movement were foremost in promoting'. The essential element was that technical teaching should be practical; more practical in fact than was normally the case on the Continent. Magnus had found that Continental education was a shade too theoretical for British tastes. The teaching of practical physics, for example, was attempted but slightly at that time in German *Gymnasia* which were still dominated by classical studies. Whether the British emphasis on the practical was, in the long run, rightly judged is something that may be doubted. It could, and in some cases certainly did, lead to an unwise disregard for the theoretical and even for the scientific.

The Central Technical College at South Kensington was opened in 1884. A three-year course was prescribed; the first year to be devoted to general science, the second and third years to engineering, physical or chemical subjects and a wide range of choice was available.* The aims of the college were fourfold: it was hoped that it would suit prospective teachers of technology, would-be architects, those intending to make a career in engineering or manufacturing industry and, finally, men who having already had industrial experience wished to study the relevant sciences. In fact, of these, the teachers were regarded as the most important; the authors of the technical education movement had seen that, before scientific technology as opposed to the old rule-of-thumb apprenticeship system, could become widely diffused as an industrial practice, there would have to be a good supply of competent teachers. What they did not see so clearly, perhaps, was that teachers must have some assurance that adequate posts await them on graduation. In the year 1888, of one hundred and seventy-five regular students at the Central College only eighty-one were intending teachers, the proportion of teachers being officially described as 'disappointingly small'. As the National Association for the Promotion of Technical Education put it a few years after the foundation of the City and Guilds Institute: 'The fact is, the demand for very high class technical education has to be created as well as the supply, and until technical classes are more widely diffused throughout the country than at present, there is little demand for the training of technical teachers'. This, it was hoped, would cure itself in time; but the members of the National Association were prepared to accept that, as for scientifically trained managers and employers, 'they will not be many for some years to come'.

A SECOND ROYAL COMMISSION

In 1880, A. J. Mundella, a firm friend of technical education, became Vice-President of the Committee of the Council. In May, 1881, H. M. Felkin published a short pamphlet entitled *Education in a Saxon town*

* Hydraulics, strength of material, practical physics, electro-technology, fermentation, crystallography, oil chemistry, dye chemistry, etc.

(Chemnitz) which described the educational efficiency of Germany, particularly in regard to the textile industry. The pamphlet was highly effective in that it provoked Mundella and his friends to obtain the appointment of a Royal Commission to investigate technical education, both in England and abroad.

This Commission* (1881–4) [24] was remarkably comprehensive: its members interpreted their terms of reference so broadly as to scrutinise institutions as diverse as Edgbaston High School for Girls and the Imperial Polytechnic in Moscow; they examined north country industrialists and trades unionists as well as Italian silk-weavers and Danish agriculturalists; they inspected German universities and French écoles and they sent Sir William Mather on a tour of inquiry into American institutions.

The Commissioners found, to their surprise, that nowhere in Europe was there a system of evening instruction in science comparable to that undertaken by the Department of Science and Art and recently supplemented by the City and Guilds. But they agreed that German primary and secondary education was better than British. Continental countries had, they concluded, made remarkable progress in the last few years. In the organic chemicals industry the Germans had, from the scientific point of view, unquestionably taken the lead and in the budding electrical industry, that owed so much to British genius, they had achieved at least parity. The Commissioners noted that the masters and managers of industry on the Continent had a high standard of scientific knowledge and they expressed their belief that industrial success could not have been so achieved without first-class technical instruction, *original research* and a widespread appreciation of the value of learning and research. Nearly all the useful institutions in the United Kingdom suffered from a shortage of funds and in this respect Britain compared most unfavourably with the Continent. The Commissioners did not feel that the value of high standard scientific training was generally appreciated in England: 'The Englishman is accustomed to seek for an immediate return and has yet to learn that an extended and systematic education *up to and including the methods of original research* is now a necessary preliminary to the fullest development of industry'.†

* Samuelson was the Chairman and the other members were Roscoe, Magnus, Swire Smith, John Slagg, William Woodall and Gilbert Redgrave (Secretary).

† Second Report, Vol. 1, p. 525 (1884).

They were told that there were, in England, scarcely any 'important metallurgical works without a chemical laboratory in which the raw materials and products were daily subjected to careful analysis by trained chemists'. The permeation of trained chemists into metallurgical and chemical works was essentially the achievement of the years following 1851. In the first half of the century there had been very few trained chemists in industry. But even at the later date, chemists were employed not as applied scientists or researchers, but as routine analysts. The true industrial laboratory had not yet emerged in England; but it had appeared on the Continent. Messrs. Bindschedler and Busch, manufacturers of coal-tar colours in Basle, had, the Commissioners discovered, an extensive establishment of chemists. Under a Director were three chemists who were Heads-of-Divisions and who had charge of a number of assistant chemists, all being university or polytechnic trained. There were research laboratories and an excellent technical library. Cases were quoted to prove that experimental investigations were followed by manufacturing and commercial success. Indeed, it was said that the Swiss coal-tar colour industry owed its success, if not its origin, to the Zürich Polytechnic. There is, after all, no coalfield in Switzerland.

As for Germany, there were, in the academic year 1883–4, no fewer than fifty students doing research in organic chemistry under von Baeyer at Munich University. In industry, the Commissioners found that Martius, of Berlin, employed fourteen chemists, while at Höchst there were, exclusive of engineers and managers, some fifty-one scientific chemists and at Ludwigshafen approximately the same number. Even the industrial physics laboratory had begun to emerge at that date. Siemens-Halske, of Berlin, had a small laboratory in which experimental as well as testing work was carried out.

In the evidence given to the Commission, W. H. Perkin showed that Germany had, in 1879, produced some £2,000,000 worth of coal-tar colours and Britain some £450,000 worth. There were seventeen coal-tar colour works in Germany and five in Britain. Thus the industry which, it was confidently expected in 1862, would render Britain independent of foreign dyestuffs, an industry that had originated in England and that depended on England's great natural resource— coal—for its raw material and on England's great textile industries for its market, had been substantially lost to Germany in less than thirty years from the date of the first discovery. Germany, remarked Perkin, knew the value of well-trained chemists, although prior to the introduction of the coal-tar industry, chemists had not been employed to any great extent in German industry. The German chemist was well educated: in order to get his degree he had to have done some original research. In this country, thought Perkin, a chemical manager should have had research experience as a necessary part of his education.

The problem of the best way to promote technical education would, Colonel Donnelly believed, be solved if one essential condition was fulfilled: that the Commissioners, or some other group, '. . . should get the employers of labour in this country generally to see and fully to appreciate the value of such instruction'.* With which we may compare Huxley's observation: 'Of all the practical measures that could be taken for the advancement of technical education and scientific teaching the most important would be that employers should show that they valued it, and that they would do something for the young people who in any way distinguish themselves'.† The sentiments were admirable but hardly original; in fact, by 1884 they had become distinctly old-fashioned. Sweeping generalisations about the appreciation or otherwise that employers showed of science were of little value. During this period no detailed surveys were carried out that would throw light on the problem: apart, that is, from the Royal Commissions and Select Committees that we have discussed. Before all of these many employers had strongly pleaded the cause of technological and scientific education. Let us grant, however, that Huxley and those who agreed with him had had personal experience that many employers in certain industries were ignorant of science and of its potential value, and that there were far fewer professional scientists in England than in, say, Germany. Then the next step should have been to find out why this was so. Had they done this they would have found that scientific talent cannot be usefully employed in isolation; that a research chemist, for example, must be supported by development personnel, production engineers, technical salesmen, etc., if his work is to be effective. If the industrial scientist does not have highly qualified colleagues with cognate skills then his work will be wasted and he, himself, frustrated. To demand, as Huxley did, that employers should value science was to require that they should break the vicious circles in which they themselves were involved.

Huxley showed much more penetration when he added: 'The history of English science is extremely instructive. Whether in physics, in

* Second Report, Vol. III, p. 287.
‡ Second Report, Vol. III, p. 322.

chemistry . . . the peculiarity of English science has been that the army has been all officers. Until within the last quarter of a century there has been no rank and file.' Our men of science were, in those days, amateurs, '. . . then between 1840 and 1850 Germany began to put out her strength in science'. While, in every branch of science we had men of original capacity equal to Germans, or to anyone else, we have not, even now, '. . . *anything corresponding to the rank and file that they have in Germany*'. In that country it is a question of organisation; men can make a living as scientists; there are, for example, many teaching posts in the rapidly growing universities and in the technical colleges.

The contrast was underlined, probably unconsciously, by Captain Abney, R.E., F.R.S., of the Department of Science and Art, when he told the Commissioners that: 'The training and education of engineer officers renders them fit persons to be acting inspectors' (of the science classes). While Sir Sidney Waterlow believed that the 'need of a high school of applied science is shown by the fact that, in this country there does not at present exist any institution which is adequately supported and has all the most recent appliances for practical science teaching'.

The recommendations of the Commission were that elementary and secondary education—fields where England was conspicuously behind Continental practice—should be radically improved. There should be more liberal provision of scholarships and local authorities should be empowered to establish and maintain secondary and technical schools. They concluded by paying tribute to those who had started the technical education movement: to Lyon Playfair for his famous letter of 1867, to Bernhard Samuelson and the Parliamentary Committee and to Mr. Felkin for his pamphlet as well as to the members of the Devonshire Commission. After this the individual Commissioners toured the country at their own expense, giving lectures in many towns and cities and trying to spread the gospel of technical education.

UNIVERSITY COLLEGES—OLD AND NEW

'It is the power of liberalising the professions that distinguishes universities from technical schools'; thus Lyon Playfair addressed the graduates of St. Andrews University in 1873 [25]. The observation was not, however, original; it constituted, in fact, one of the yardsticks whereby Hamilton had measured the shortcomings of the universities in his day: maintaining that the English universities had abandoned professional teaching while the Scottish universities had failed to allow for the liberal element by concentrating too much on the professions, such as medicine and the law.

Over the period of the early technical education movement the development of university education in London and the provinces had been uneven and in many respects unsatisfactory. The healthiest development was in Manchester, where the Owens' College, having rounded the corner, commenced a period of steady expansion. Larger premises were acquired in 1868 despite the refusal of first Disraeli and then Gladstone to give any aid. Excellent new chemistry laboratories, designed in the light of the best Continental ideas, were also built. In fact Owens' Chemistry Department was particularly prosperous and in 1874 Roscoe was able to get Carl Schorlemmer appointed Professor of Organic Chemistry—an unprecedented step in England, although of course some twenty-one years after the chair of organic chemistry had been established in Paris. But Roscoe went even further and instituted a four-year course in applied chemistry, with prescribed classes in mathematics, physics, engineering, drawing and French or German as well as in the standard chemical subjects.

It was, therefore, natural that Owens' College, although still short of money, should aspire to university status. To this end Roscoe and his colleagues worked hard, enlisting considerable outside support,* but at the same time incurring the hostility of Robert Lowe [26]. To grant university status to Owens' would be seriously to jeopardise the accepted standard of degrees, for what better guarantee of a respectable standard can there be than strict examination conducted by an outside authority? Without an omnipotent and omniscient examining board the almost certain result must be a Dutch auction of degrees. So, at any rate, reasoned Lowe who was always true to the doctrines of individualism. Fortunately, however, he did not prevail and the Victoria University Charter was granted in 1880. This, in its original form, was a federal university, comprising the Owens' College, University College, Liverpool, and the Yorkshire College of Science (now Leeds University). It was not, like London University, a mere examining board, for it was entirely controlled by the constituent colleges and there were no 'external' degrees.

In London, University College had its Department of Chemical Technology, Charles Graham having been appointed to the Chair in

* Among the supporters were Kelvin, Sir B. C. Brodie, jun., Lyon Playfair, Mark Pattison and T. H. Huxley.

1878. The variety of subjects was very wide and included metallurgy, brewing and breadmaking. These, however, were short, self-contained lecture courses and there was no extended course in applied chemistry like the one Roscoe had organised at Manchester. Nevertheless the college had no reason to regret its Chemistry Department which was one of the most active of all: the chair was occupied by A. W. Williamson, one of the abler English chemists of the later nineteenth century.

Another interesting innovation was the teaching of practical physics. Although Kelvin was the first to open his laboratory to students the introduction of systematic practical physics tuition was due to W. Grylls Adams at King's College, London, and R. B. Clifton at Oxford. Cambridge followed suit in 1872 and University College, London, in 1876.

The teaching of practical physics necessitates the establishment of laboratories. At Cambridge and Oxford the foundation of the Cavendish and Clarendon laboratories were events of historical importance; they indicated the full recognition of the 'progressive' sciences by the universities; no longer was the scene to be dominated by the old mathematics and natural philosophy disciplines.* The reform, obvious as it now seems, did not meet with unanimous approval at the time. It was criticised, for example, by that distinguished Cambridge scholar Dr. Isaac Todhunter who expressed the conviction that the student should be prepared to accept whatever his master told him. For holding this view Todhunter has been frequently, and often unfairly, criticised by later reformers. His opposition to 'experimental' studies was not obscurantist but, on the contrary, can be shown to be, in one respect, well founded. In Todhunter's opinion: 'Experimental science, viewed in connection with education, rejoices in a name which is unfairly expressive. A real experiment is a very valuable product of the mind, requiring great knowledge to invent it and great ingenuity to carry it through.' Furthermore, the experimenter, 'like the poet, is born and not manufactured'.† Now this opinion of Todhunter's calls for one comment only: it is evidently perfectly correct and it enshrines a truth that is too often conveniently forgotten, even to-day. Quite obviously, a class 'experiment' differs fundamentally from that activity

---

* Yet the expansion of the Cavendish Laboratory was very slow; it took more than forty years for the number of research workers to rise to about twenty five. This was in spite of the famous men who were successively Directors of the Laboratory: Maxwell, Rayleigh, Dewar, J. J. Thomson.

† Isaac Todhunter, *The conflict of studies and other essays* (London 1873).

which Faraday was referring to when he wrote his *Experimental Researches*.

In one sense, however, Todhunter was wrong, for the foundation of these two laboratories marked the beginning of genuine research and experiment in university physics. Had not Maxwell already distinguished between 'experiments' of illustration and experiments of research? Nor was research confined to the two famous laboratories; at University College, London, for example, students in the laboratory were required to make their own apparatus and to carry out their own original investigations [27]. In other words, there was a degree of freedom of learning; for the students of those days, other than the candidates for the external London degrees, were not under immediate examination pressure.

While Owens' College and University College were comparatively prosperous the states of the two ecclesiastical foundations were much less secure. Durham University was extremely poor and a school of physical science, founded in 1865, proved a complete failure [28]. Not until 1871, when the Armstrong College of Science was established, did the prospects for science look up at that university. King's College, London, was in desperate straits, being reduced in 1875 to selling its spoons.* The Rev. Alfred Barry, appointed Principal in 1868, was well disposed to the scientific movement and made efforts to build up the technological departments of the college. Physics and engineering were developed with the aid of the City and Guilds and the award of Whitworth Scholarships. But it was not enough and the main source of revenue continued to be the popular evening classes.

A very important movement to which the supporters of technical education contributed was the creation of those provincial university colleges that were founded in a number of large cities from about 1870 onwards. One of the first instances was in the north when a number of industrialists persuaded Durham University to grant £1,000 a year for six years to aid the establishment of a science college. An endowment appeal was very successful and in 1871 Armstrong College was opened; its purpose was to supply specialised instruction in mining and engineering together with courses in mathematics, physics, chemistry, geology and biology. In 1873 Josiah Mason endowed the Birmingham Science College, and in October 1874 the Yorkshire College of Science was opened, the Clothworkers' Company endowing eight scholarships in textile technology there. Two years later the

* F. J. C. Hearnshaw, op. cit.

University College of Nottingham was opened, a municipal institution that enjoyed a subsidy from the rates. Not long afterwards Firth College, Sheffield, and University College, Liverpool, were founded. Within a short time many of these new, liberal science colleges were teaching a large number of purely technical subjects: plumbing, carriage making, hosiery, etc. They were, in their early youth, rather like the more liberal technical colleges of the present day.

Besides the technical education movement, the 1870s witnessed the remarkable development of the university extension movement. As long ago as 1850, the Rev. William Sewell of Exeter College, Oxford, had suggested the formation of university lectureships in certain industrial towns. With the flood tide of university reform and the repeal, *in toto*, of the Test Acts, this suggestion became a real possibility. In 1871, James Stuart, of Trinity College, Cambridge, was able to get official approval for the idea and before very long university enthusiasts were giving lectures to classes of artisans in the industrial areas. This led directly to the foundation of the University of Bristol (1873); Balliol College and New College, Oxford, each contributed £300 a year for five years on the understanding that the education provided would be liberal .

The later history of the university extension movement is outside the scope of this account. It is hardly relevant to science and technology; it belongs more to the field of activities of such organisations as the Workers' Educational Association. No one would wish to denigrate the services rendered by this movement nor the selfless motives of the young university men who did so much to forward it in the early days. But it is impossible to accept such judgments as that of Lewis Campbell: 'The provincial universities are the outcome of the extension movement and have, in part, supplanted it. . .' . [29]. This does much less than justice to the founders of Owens' College, the technical education movement, the Society of Arts, Lyon Playfair and many other individuals whose efforts were directed at precisely the same objective and who, with the greatest respect for the universities, did more and worked for a longer time to achieve it.

There were two other noteworthy innovations during this decade. In 1872 the Department of Science and Art constituted the so-called day organised science schools. These were secondary schools that were eligible for grants for teaching science, either as a whole or for particular classes. Several of the reorganised endowed grammar schools took advantage of this to gain State support for their science classes.

The second innovation occurred nine years later when, in 1881, the Treasury approved a Minute proposing the reorganisation of the South Kensington colleges as a 'Normal School'; that is, as a college whose main purpose was to train teachers.

A link between these two events is indicated by the work of Edward Frankland.* As Professor of Chemistry at South Kensington he had become exasperated by the evident ignorance of practical chemistry evinced by the examinees who had been trained under the payment-by-results system (see p. 129 above). In 1869, therefore, Frankland inaugurated the annual South Kensington summer schools for teachers, the aim of which was to raise the standard of chemistry teaching, particularly on the experimental side. He was also able to persuade the Department of Science and Art to provide grants for school laboratories. The subsequent development of practical training and in particular the 'heuristic' system advocated by H. E. Armstrong owe a great deal to Frankland's work at South Kensington.

ARNOLD AND PATTISON

It is a commonplace that the intellectual climate at Oxford during the first half of the nineteenth century was cold, measured by the scale of science and scientific education. This did not, however, deter substantial numbers of Oxford men from being keen advocates of science, both as a mode of education and as an intellectual endeavour; and it was, perhaps, no accident that, when the reform movement gathered strength in the 1860s, two Oxford scholars should publish, in 1868, books that revealed very clearly the trend of educational thought of a large and influential group and that were destined to leave their marks on the later course of academic reform. The first of these books was written by Matthew Arnold,† the son of the educationist and himself an experienced school inspector as well as a poet. Arnold's theme was the schools and universities of Germany and the lessons that England could learn from them. Not that he was the first to call attention to the excellence of the German educational system—Walter Perry, for one, had done that many years before—but Arnold's fame, his literary abilities and his mastery of his thesis gave his book an authority that could hardly be rivalled.

* I am grateful to Dr. W. H. Brock for the information on which this paragraph is based.
† *Higher Schools and Universities in Germany* (London 1868).

According to Arnold's analysis, the professional half of the English middle class—clergymen, lawyers, doctors—had usually been educated with, and so carried the cachet of, the aristocracy. The young men of this class, of the great public schools and universities, had fine governing qualities but lacked an understanding of science: '. . . having an indisposition and incapacity for science, for systematic knowledge'. In contrast, the other half, the commercial and industrial half, had been educated in a way that was an inferior copy of the first (cf. Whewell's comments) and this was disastrous, for failing in the compensatory social qualities, they also necessarily lacked science. On the Continent it was quite otherwise: the nations' middle classes were the strongholds of science. In England the workers were uneducated, the middle classes educated on the second plane and the idea of science absent from the whole structure and design of education.

It followed from this lack of scientific education that reliance on rule-of-thumb was the characteristic English, as compared with German, industrial practice; even in the engineering industries, where one would have supposed that England led the world, there prevailed a miserable system of 'blunder and plunder'. More than this, '. . . we hardly even know the meaning of the word science in its strict sense and employ it in a secondary and incorrect sense'. It is in science that England needs to borrow most from the German universities: 'The French universities have no liberty and the English universities have no science; the Germans have both'.

Arnold knew that the universities and technical high schools of Germany were the crowns of a long, co-ordered series of educational establishment: primary schools, trade schools, *Realschülen, Gymnasia*, designed, graded and evolved by some of the best brains in the country. He had, too, clear ideas about the functions of university and technical college; in the former the idea of science predominates, in the latter it is subordinate to professional requirements. Unfortunately, in England neither Oxford nor Cambridge fostered the idea of science; with their collegiate and tutorial systems and with their written examinations and degree requirements they were not really universities but merely high schools—*hauts lycées*. They did not take education beyond the general school level. As for London University, it was no university at all; it was merely an examining board. In Germany on the other hand, the doctorate degree necessitated a thesis, evidence of original work, and in no German university was 'the degree examination, in itself, such as to make it what the degree examination is with us . . . the grand, final cause of university life'. Examinations did not have a significant rôle in German universities; and, he added, '. . . examinations may be a protection from something worse. All I say is that a love for things of the mind is what we want and that examinations will never give it to us.'

The eighteen fifties and sixties were the years when enthusiasm for written examinations stood at its highest. Arnold's criticisms are, therefore, of great interest. Apart from stubborn reactionaries and traditionalists of the old school, he was one of the first to have reacted against the new system in so far as it impinged on higher education. In later years criticisms of higher examinations, indeed even of elementary examinations, became much more common.

What would Arnold recommend? The 'University' of London must become a teaching body and to make this possible Oxford and Cambridge must each donate a share of its teaching staff. Even more daring, and to ensure the success of these reforms, there must be a system of true public secondary schools, a Ministry of Education and a redistribution of endowments. As for the proper form of scientific education, Arnold felt that it should be such as to arouse interest in the entire circle of knowledge; but human faculties are limited and 'it is for the most part through a single aptitude or group of aptitudes that each individual will get his access to intellectual life and vital knowledge, and it is by effectually directing these aptitudes on definite points of the circle that he will really obtain his comprehension of the whole'. Under our present educational arrangements we are rapidly losing ground; while 'our disbelief in government makes us slow to organise government for any matter'.

The second important book to be published in 1868 was written by the Rev. Mark Pattison, Rector of Lincoln College.* Unlike Arnold, Pattison had spent his whole adult life at Oxford and was, therefore, very familiar with the changing nature of the university during the critical years of reform. Indeed, Pattison's intellectual odyssey was a microcosm of university history; as a young man he had succumbed to the appeal of the Newmanites but later he broke with the Tractarians and, by 1851, was pleading the cause of university reform before the Oxford Commissioners; later still he moved further to the intellectual left, becoming markedly rationalistic in outlook. Accordingly, Pattison's *Academical Organisation* proposed the application of much the same principles as those advocated by Matthew Arnold to the reform of

* *Suggestions on Academical Organisation* (London 1868).

Oxford University. His work also resembled Arnold's in that it was very widely read and its ultimate importance in determining the course of university reform has been judged by A. I. Tillyard as comparable with the writings of Hamilton [30].

For Pattison the most important thing is that endowments should be redistributed in such a way as to convert Oxford into a scientific institution. The despised pass degree must be abolished and the Master of Arts degree conferred immediately on graduation. In addition the mode and method of instruction must be the 'exclusive devotion of the mind to some one branch of science'. This was a sanction for specialism for which Pattison did not think it necessary to give reasons, for these 'would involve our mental constitution and a survey of the history of universities either of which are much beyond my present limit'.* But the phrase 'our mental constitution' can be taken to mean almost anything and as universities necessarily reflect the societies in which they have their being it is difficult to see what, in this respect, can be proved by a survey of their history. More significantly, perhaps, he speaks of 'the external pressure of a profession' and adds that 'the division of labour is the law of mental, no less than of manufacturing production'; elsewhere he talks of this as being 'in the reason of the thing'.

Only in the matter of research was Pattison out of step with Arnold. Research, for Pattison, was, at that time, subordinate (although he later changed his mind). Possibly he still thought that the main scientific effort would continue to be outside the universities and in the hands of the large, quasi-amateur groups of doctors, lawyers, clergymen, army officers, engineers, etc. who continued to form the greater part of the body scientific in England. In other respects his views were much the same as Arnold's. The university should be an association of professional scientific men whose duty was to maintain, cultivate and diffuse extant knowledge, while from the students' point of view the aim was to be specialised instruction in the sciences. And Pattison leaves us in no doubt about this: quite logically he criticises London University for the broadness of the B.Sc. syllabus†—a flaw that was, however, 'redeemed and to some extent cured at the next stage', for the D.Sc. syllabus required a 'thorough and scientific study of one branch of science'.‡

* Pattison, op. cit., p. 262.
† Pattison was examiner in logic to the University of London.
‡ Pattison, op. cit., pp. 276–7. Despite this illiberal philosophy of higher

Pattison wanted the universities to be national and open to all the talents; no longer must they be the preserves of the wealthy. More, he foresaw the end of the old *laissez-faire* individualism, for 'civilisation in the west has now reached a point where no further triumphs await mere vigour undirected by knowledge. Energy will be beaten in the practical field by combined skill . . . the conviction must, ere long, reach us that our knowledge is defective and such is the length of art and the shortness of life that knowledge can only be made available for public purposes by concert and organisation.'*

Opinions of this sort were, of course, of great significance, coming as they did from an influential and respected Oxford scholar. It is quite apparent that with Pattison we are in a very different social and intellectual atmosphere from those familiar to Whewell and, *a fortiori*, to Newman. The clear demands for an open university and for social justice in education were great advances on Whewell's complacent acceptance of the *status quo*. On the other hand, for many, Newman's apology for the old Oxford and his gentle liberalism are not without great appeal, even today.

THE PATTERN OF EXAMINATIONS

In advocating specialised study Arnold and Pattison were approving a trend that had been apparent for some time. The Mathematics Tripos was already specialised and when, in 1851, the Classics Tripos was freed from mathematical obligations this, while hailed as a reform, was, as Todhunter remarked, also a step towards specialisation. It would have been possible to have met the just complaints of the classicists by demanding some degree of literary culture from the mathematicians as well as some degree of mathematical learning from the classicists.

Experience shows that increasing specialisation is characteristic of written examinations: educational authority tends to 'rationalise' them. Mathematics examinations, the first to be specialised, permit of more accurate marking and classification than, probably, do any others.†
Consequently, as Henry Latham pointed out,‡ extraneous subjects in

---

education, Pattison was very much alive to such examination abuses as 'cramming'; he held that examinations alone, without proper teaching, were 'unwholesome'.

* Pattison, op. cit., p. 329.

† A calculation can only be correct or incorrect. An essay, on the other hand, cannot be appraised in such a simple and objective fashion.

‡ *On the action of examinations considered as a means of selection* (London 1877).

O.S.E.—6

a mathematics examination would tend to destroy the homogeneity of the course. For example, 'the slight study of moral philosophy' was dropped from the Mathematics Tripos as early as 1828 for it was felt that, as between two would-be Senior Wranglers, it would be most unjust to allow competence in moral philosophy to decide the issue. The same consideration was apparent when the Oxford Public Examiners submitted, in 1831, that logic 'be not absolutely required of candidates for Mathematical Honours' [31], because: 'The requisition put upon candidates for Mathematical Honours to pass in logic acts as a direct discouragement to Mathematical studies, besides being in itself unjust, as the candidate has not the option allowed to others in the general examination, but is obliged to undergo an examination in logic, in addition to Mathematics'.

In a similar way the oral examination, once a feature of the Mathematics Tripos, had quietly disappeared. Whewell had approved of oral examinations; conducted by competent examiners there was no better way of exploring the highways and byways of a candidates knowledge, no surer method of unmasking the mere crammer. But unfortunately it is very difficult to conduct an oral examination in such a way that all candidates get exactly the same 'value' of questions. To submit the same questions to all candidates would, on the other hand, be to sacrifice flexibility, the main virtue of the oral examination.

However strong is the belief that it is desirable for students to acquire a truly liberal education, liberal in content as well as in intent, if for the above reasons, it is impossible to include extraneous subjects in the examination papers, all that can possibly be done is to allow time and opportunity for their pursuit in the hope that students will seek voluntarily to broaden their education. But this would almost inevitably result in lowering the degree standard and would, in the event, give *carte blanche* to the indolent. If a respectable degree standard is to be achieved then the course must be such as to exercise the student to the full limit of his abilities. If this is not the case—well, there were few, if any, defenders of the pass degree among nineteenth-century reformers.

The process of rationalisation was also apparent in the case of the Natural Sciences Tripos. This had originally consisted of examinations in five science subjects (see p. 97) and although some candidates restricted themselves to one or two subjects, a number, in the hopes of gaining more marks, attempted several. This, it was felt, was leading to superficiality and so, from 1874 onwards, attempts were made to ensure that the courses encouraged 'thorough knowledge of at least one subject—so far as it can be mastered during the period of under-graduate study—combined with a sound study of the principles and leading facts of the cognate and subsidiary subjects. . .' [*32*]. This led to a division of the Tripos into two parts. The first part, corresponding to the pass degree, required the elementary portions of three or more subjects, while in the second part only one subject was important although a competent knowledge of one other was required. The first part could be taken at the end of the second year; the second part at the end of the fourth year.

At the same time that these modifications were being made to the Cambridge science degrees, the London University was following exactly the same course. The original London B.Sc. had been carefully designed to ensure that the candidates possessed '. . . such general culture as should be likely to prevent its holder from being a mere specialist'. But, on 27 May, 1876, a committee of the university announced that '. . . a thorough knowledge, limited to a comparatively small range, is preferable to a slighter acquaintance spread over a more extended area. And it is the general experience of teachers that there is, from the commencement of their academical course, such a decided preference on the part of nearly all students of science for either the physical or the biological group of subjects, that the attention of each student is given to one group almost to the exclusion of the other.' To which they added: 'Teachers and examiners assure us that this is best' [*33*]. The Honours regulations were, of course, left unchanged.

These developments clearly ran counter to the definitive recom-mendations of the Devonshire Commission and to the express opinions of many of the leading scientists of the day (see above, pp. 94, 123, 125). Apart from the advocacy of men like Pattison—and they interpreted a trend rather than suggested a reform—what reasons can be found to account for this precipitate abandonment of the idea of a comprehensive and liberal education?

Why, that is, do examinations become specialised?

In the first instance we can consider the general climate of opinion. The doctrine of the division of labour had transcended the works of the economists and had become a Victorian article of faith, enjoying almost metaphysical status. Newman acknowledged its force; Spencer found it to be an ingredient of the 'law' of progress [*34*]; the Prince Consort gave it his Royal approval [*35*] and it was only natural that before long it should be applied to the world of learning. The argument was very simple: a man who devotes his attention to one object is more likely to effect an improvement in that object than one who allows his interest wider scope.*

A more tangible reason, and one explicitly given, was the impressive advance of knowledge during the nineteenth century. The 50 years preceding 1881 were marked by great advances in natural science and it is certain that these alone would necessitate some degree of specialisation. But, on the other hand, these advances were mainly steps in the direction of greater unification: of that greater generality which is the hallmark of scientific progress. Correlated together were heat and energy [36]; light and electromagnetism; physics and chemistry; organic and inorganic chemistry, etc. In a real sense, then, science in acquiring greater generality becomes simpler; the phenomena are not so diverse and perplexing. Again, it cannot be said that all knowledge is of equal value and the question of value can only be judged in the light of a critical philosophy of science and beyond that of a social philosophy, a consideration of the ends involved. It was an unfortunate feature of the examination system as it was operated that it tended to ignore such considerations; utility for examination purposes, as Todhunter saw clearly, rather than intrinsic importance being a criterion for the selection of questions. Thus the important features of a science tended to be submerged in a sea of minor detail and the subject to become unnaturally expanded in consequence.

Demand for applied scientists cannot have been a factor in the specialisation of studies, for industry at that time offered extremely few opportunities for the highly trained chemist and even fewer for mathematicians and physicists. At Cambridge, for example, the institution of highly specialised courses resulted in a marked and continued fall in the numbers taking the second part of the Natural Sciences Tripos and a rapid rise in those taking the first part, the pass degree (see p. 205). Also, were it a question of economic forces, of demand and supply, we should hardly expect them to be manifest at Cambridge, whose alumni were still largely young men of comparative means. Professional science can, then, be ruled out as a factor leading in the first instance to specialisation. In one profession only—that of teaching—would the specialised graduate be in some demand. The public schools would value the graduate with high Honours not merely for his prestige but also for his ability to coach boys for the university

* J. S. Mill expressly warned against the application of this theory to invention, etc.

and college scholarships. Scholarship examinations had also become highly specialised at an early date and the reputation of the school rested to some extent on the number of scholarship winners it could boast.

The trend towards specialisation was not limited, of course, to mathematics and science. With the specialisation of the Natural Sciences Tripos there went, *pari passu*, the specialisation of the Classics Tripos and this process rapidly became general for other subjects as they were introduced into university syllabuses. Evidently the proliferation of specialisation is accounted for rather by the very nature of academical organisation than by any concrete external demand. Examination stimulates study and, as the London examiners pointed out, the student who usually has a preference, or aptitude for one branch of knowledge, would naturally prefer to be examined solely in that branch than be compelled to study other branches as well. Therefore under the conditions imposed by written examinations specialisation is agreeable to the examinee, generally speaking, as well as to the examiner. And it is true that any advanced study must, of necessity, be specialised whether it is stimulated by written examinations, thesis requirements or simply the desire to advance knowledge. But in the case of free research the specialism is voluntarily selected by the student; in the case of the set syllabus and the written examination it is imposed from without by the teachers and is governed by past experience of other students.

The machinery for high-grade specialisation had, therefore, been assembled and perfected by 1881; before there was any noticeable demand for professional scientists. The main reasons for it seem to be in the nature of the teaching and examining systems together with the obvious existence of markedly different interests and aptitudes among the students.

CRITICISMS AND SUGGESTIONS

Specialisation in higher education, the isolation of one science from another and science as a whole from the humanities, has always evoked criticism and it would be wrong to suppose that the developments of the seventies and eighties passed without comment. In 1885 Lyon Playfair, speaking as President of the British Association, explicitly condemned specialism: '. . . the divorce of culture and science which the present state of education in this country tends to produce, is deeply to be deplored because a cultured intelligence adds greatly to the development of the scientific faculty'.

With much less precedent, however, the instrument of specialisation, the written examination, also began to suffer criticism. Arnold was, as we saw, no lover of examinations and he was only the first of many critics who emerged during the last decades of the century. Todhunter, an original mathematician as well as a teacher and examiner of experience, asserted as early as 1873 that 'The excessive cultivation for examination purposes of one department of knowledge to the exclusion of others seems to me to be one of the great evils of our modern system of bribing students by great prizes and rewards to go through our competitive struggles'.* He was emphatic that specialisation was a 'serious evil' and he hoped for a more general culture than it allowed. He was, therefore, quite consistent in his desire, strange as it seems, that mathematical examinations should be both shorter and easier and he wished to '. . . join my protest, feeble as it may be, with that of other persons both within and without the university against the exorbitant development of the system of competitive examinations'.†

In the previous year Lyon Playfair, criticising a characteristic suggestion of Robert Lowe's, could say to an academic audience and with the subsequent approval of many academic authorities: '. . . one great evil of university education, and more particularly of university examination, is to create faithful disciples rather than independent thinkers' [37]. In case it is thought that this is unfair to Lowe, let it be remembered that in the opinion of the latter the Civil Service Commission could well be expanded to perform the function of a 'university', for examining is a judicial function [38].

It is, of course, one thing to criticise; it is usually quite another to suggest remedies. But in this case there was an alternative in the field; for the most part those who criticised examinations were either men who had been educated at German universities or were avowed admirers of German education. Thus, in 1872, a witty individual, a 'British and Foreign Graduate', praised the German Ph.D. system and denounced the English practice on the grounds that examinations are designed, not to discover the intellectual development and mental standards of a candidate, but to find out how much he has learnt from other men; examinations suit the acquisitive mind, while the thesis system suits the creative mind [39]. In the following year Roscoe, also

* Todhunter, op. cit., p. 5.
† Ibid, p. 241.

educated in Germany, was stressing the vital importance of research: '. . . on the Continent of Europe, to a great extent, and in the United States, in some measure, those who wield the sceptre of government are not only aware of the national importance of original research, but, what is more, they act up to their convictions whilst we feel that the same cannot be said of our country.' The essence of the scientific spirit is that it is free and disinterested, that it will not be bound by any constraints, and for these reasons research must be recognised as a means of education. Indeed, Kolbe had told Roscoe that German chemical manufacturers were refusing to employ young men unless they had had previous research experience. In England the universities ignored the claim of research and 'Sir William Thomson (Kelvin) had expressed his opinion that the system of examinations at the university has a tendency to repress original enquiry, and exerts a very injurious effect in obstructing the progress of science'. Germany has practised *Lehr- und Lernfreiheit* but 'if the students had been all ground down to the one pattern by the requirements of competitive examinations, originality of mind would have been effectually discouraged' [40]. Kelvin, incidentally, had told the Devonshire Commission that, in his opinion, examination exerted a 'fatally injurious tendency' on the higher parts of science.*

Huxley, examiner though he was, questioned the value of university examinations in his Rectorial Address at Aberdeen University in 1874 [41]. He agreed that written examinations tell us next to nothing about a man's powers as an investigator and conceded that there was much to be said on behalf of research degrees—a practice which, he reminded his audience, was of some antiquity. Three years later he went so far as to describe competitive examinations as 'The educational abomination of desolation of the present day . . .'. In that year (1877) Walter Perry was asking whether 'cramming' and examinations are the best means of enlarging the mind; for it was his belief that 'the mind grows best on the food that it chooses for itself' [42], and it was his fear that private tuition and regulated study would destroy the scientific spirit.

That professional, or industrial requirements were not among the causes of the specialisation of science degrees follows from J. Norman Lockyer's observation, 'There is absolutely no career for the student of science as such in this country. True scientific research is absolutely unencouraged and unpaid' [43] (1873). Not that Lockyer was the only one to deplore this, for there was at that time a small group actively

* See also Vol. II of S. P. Thompson's biography of Lord Kelvin.

trying to remedy the situation and also to reduce the importance of examinations in the university. The pioneer of the group was Dr. C. E. C. B. Appleton, a young philosophy lecturer at Oxford who, after graduation, had spent a couple of years at German universities. Appleton returned home full of enthusiasm for *Wissenschaft* and, the times being propitious, was able easily to persuade a number of younger university men of the value of research and of its true position, or rather, what should be its true position, in the university.

Stimulated by Appleton's enthusiasm a group of distinguished academics and others interested in the problem met* for a discussion at the Freemason's Tavern. Mark Pattison took the chair and the first speaker, Professor George Rolleston, after a bitter denunciation of the examination system—'Men get demoralised by the process'—went on to ask how best to encourage the few to do research without sacrificing the interests of the mediocre? After W. B. Carpenter and Burdon Saunderson had compared the English universities unfavourably with the German ones, Pattison proposed the first motion: 'That to have a class of men whose lives are devoted to research is a national object'. Having carried the motion unanimously the company agreed that the best means of achieving this would be by redistributing university endowments in favour of active research workers. Pattison cautioned against an all-out attack on the practice of examinations—'however much we may be convinced of its effect . . . as actually carried out, in sacrificing literary and scientific ability'—and proceeded to recommend as the major objective, the 'planting' of research professorships in the academic body. It was thereupon resolved to establish a 'Society for the Organisation of Academical Study' and a provisional committee was elected [*44*].

A month later, in December, a letter from Sir B. C. Brodie appeared in *Nature*, calling attention to the proposed Society and asserting that 'Science requires the services of a class devoted to the extension of knowledge, precisely as other classes are devoted to commerce, to politics, or to agriculture. Such a class does not exist among us and its absence is the greatest defect in our social system'. Unfortunately our universities are now superior kinds of 'grammar schools' and 'of the university as thus understood, pecuniary prizes were to be the motive power, and competitive examination the regulating principle'. Brodie ended with a strong appeal for membership of the Society [*45*].

A few years later Appleton went on to publish, some nine months

* In November 1872.

before the appearance of the Eighth Report of the Devonshire Commission, a draft scheme for the endowment of research. The ideas he put forward were similar to those propounded in the Eighth Report and adopted by the government (see p. 125) [46]. But the Society he had inspired ultimately failed, in spite of the interest of men like Lockyer and Roscoe. Partly this was because the universities were at that time much concerned with the extension movement.

Before we return to a consideration of the technical education movement it is interesting to record two recantations by former advocates of the examination system. Towards the end of 1870, James Booth wrote to the *Journal of the Society of Arts*, claiming to have originated examinations in the public sphere, but adding that excessive examination 'now so common, is much to be deplored' [47]. And Mark Pattison, in his *Memoirs* [48], published posthumously in 1885, wrote of the period of examinations and university reform: 'Little did we foresee that we were only giving another turn to the examination screw, which has been turned several times since till it has become an instrument of mere torture which has made education impossible and crushed the very desire for learning'.

REFERENCES

[1] *J.S.A.*, xv, 7 June, 1867.

[2] *The Times*, 29 May, 1867 et seq.

[3] *J.S.A.*, xvi, 31 January, 1868.

[4] Ibid., 24 July, 1868.

[5] *Hansard*.

[6] *Report of the Select Committee on Scientific Instruction ordered, by the House of Commons, to be printed* (London, July 1868).

[7] *Report of the Board of Education for the year 1913–1914*, Cd. 7934.

[8] *J.S.A.*, xxiv, 11 July, 1873.

[9] For an account of government aided scientific institutions in Dublin at this time, see B. B. Kelham, *Science education in Scotland and Ireland, 1750–1900.*

[10] Lyon Playfair, 'On Primary and Technical Education', two lectures to Edinburgh Philosophical Institution (Edinburgh 1870).

[11] Col. Alexander Strange, *Report to the 1868 Meeting of the B.A.* (Mathematics and Physics Section), p. 6.

[12] *Report of the 1869 Meeting of the B.A. at Exeter*, p. 213.

[13] Col. Alexander Strange, *J.S.A.*, xviii, 8 April, 1870.

[*14*] Col. Alexander Strange, *Journal of the Royal United Services Institution* (15 May, 1870) no. 64.

[*15*] Col. Alexander Strange, *Report to the 1871 Meeting of the B.A.* (Mathematics and Physics Section).

[*16*] *Report of the Royal Commission on scientific instruction and the advancement of science* (4 vols., London 1872–5).

[*17*] R. M. McLeod, 'The Support of Victorian Science: the Endowment of Research Movement in Great Britain, 1868–1900', *Science and Society*, edited by Peter Mathias (Cambridge University Press, 1972) pp. 197–230.

[*18*] *J.S.A.*, xvii, 26 March, 1869. An interesting example of the relationship between the State and science is discussed by R. M. McLeod, 'Science and Government in Victorian England: Lighthouse Illumination and the Board of Trade, 1866–1886', *Isis* (Spring, 1969) lx, pp. 5–38.

[*19*] *J.S.A.*, xx, 29 December, 1871.

[*20*] Ibid, lxxvi, 18 November, 1927.

[*21*] Ibid, xx, 23 February, 1872.

[*22*] Ibid, xx, 26 July, 1872.

[*23*] Philip Magnus, *Educational Aims and Efforts, 1880–1910* (London 1910).

[*24*] *Report of the Royal Commission on Technical Instruction* (4 vols., London, 1882–4).

[*25*] Lyon Playfair, 'Universities and Professional Education', an address at St. Andrews University, 8 February, 1873. Printed in *Subjects of Social Welfare*.

[*26*] Robert Lowe, *Fortnightly Review* (February, 1877) pp. 160–71.

[*27*] *Calendar*, University College, London (1873).

[*28*] C. R. Whiting, *The University of Durham, 1832–1932* (Sheldon Press, 1932) pp. 113–17.

[*29*] Lewis Campbell, *The Nationalisation of the Old English Universities* (London 1901) p. 235.

[*30*] A. I. Tillyard, *A History of University Reform, from 1800 to the Present Time* (Heffers, Cambridge 1913).

[*31*] Papers submitted to the Vice-Chancellor of Oxford and the Heads of Houses by the Public Examiners, 1 December, 1831.

[*32*] Silvanus P. Thompson, *Life of Lord Kelvin*, ii, p. 1132. See also W. W. Rouse Ball, *A History of the Study of Mathematics at Cambridge*, p. 216 and D. A. Winstanley, *Later Victorian Cambridge* (Cambridge University Press, 1947).

[*33*] *Minutes of Committee of the University of London, 1867–1880*.

[*34*] Herbert Spencer, 'Progress: Its Law and its Cause', *Westminster Review* (April, 1857). Reprinted in *Essays on Education*, (Everyman edition, London 1949) p. 153.

[*35*] Prince Consort, Presidential address to the 1859 meeting of the B.A. at Aberdeen.

[*36*] See, for example, D. S. L. Cardwell, *From Watt to Clausius, the Rise of Thermodynamics in the Early Industrial Age*.

[*37*] Lyon Playfair, 'On Teaching Universities and Examining Boards', an address at Edinburgh University (Edinburgh 1872). Also printed in *Subjects of Social Welfare*.

[*38*] A. Patchett Martin, *The Life and Letters of the Rt. Honourable Robert Lowe*, i, p. 30.

[*39*] 'British and Foreign Graduate', *Doctors Dissected: or University Degrees Fairly Represented* (London 1872).

[*40*] H. E. Roscoe, 'Original Research as a Means of Education', *Essays and Addresses by the Staff of Owens' College* (Manchester 1874).

[*41*] T. H. Huxley, 'Universities: Actual and Ideal'. Address at Aberdeen University (Aberdeen 1874).

[*42*] W. C. Perry, 'On German Universities', *Macmillan's Magazine* (December 1877) xxxvii, no. 218.

[*43*] J. N. Lockyer, 'The Endowment of Research' (1873), printed in *Education and National Progress*.

[*44*] *Nature* (28 November, 1872) vii, pp. 72–5.

[*45*] Ibid (12 December, 1872), vii, pp. 97–8.

[*46*] R. M. McLeod, op. cit. [*17*].

[*47*] *J.S.A.*, xix, 16 December, 1870.

[*48*] Mark Pattison, *Memoirs* (London 1885) p. 303.

CHAPTER SIX

The Years of Failure—
and of Reform:
1888–1902

The last decade of the nineteenth century was an intellectual epoch later to be described as one of the dullest stages of thought since the First Crusade: 'The period was efficient, dull and half-hearted. It celebrated the triumph of the professional man.'* In England the old free amateurism was disappearing and with it those great figures, men like Darwin and Joule, who had contributed so much to the scientific achievements of the century. The new professional scientists, graduates of the recently introduced natural sciences schools of the older universities and of the budding provincial universities, were as yet few in number and, one suspects, inferior in the quality of their training to those coming from the German universities, then at the pinnacle of their fame.

Pedestrian the age might have been but, as so often in the past, the torpor proved to be the calm before a series of massive upheavals that were later to transform the physical sciences: the quantum theory and relativity, which were essentially German in origin, and the study of radio-activity which began in France with the work of Becquerel and the Curies. If, in England, the contemporaneous discovery of the inert gases by William Ramsay and the speculations about the entity that Stoney called at first the 'electrine' and then later the 'electron' were less immediately fundamental than the Continental developments, an optimist might still find cause for modest satisfaction. After all, in the forty years or so before Ramsay's discoveries chemistry in Britain had almost died on its feet.

It is true that young Germans and Swiss had often been employed in England in occupations for which their technical training was superior to that available to young Englishmen. In this way the great locomotive building firm of Beyer-Peacock, founded in 1854, employed a German as head draughtsman almost from the beginning and usually

* A. N. Whitehead.

had a number of other Germans in the drawing office [1]. This was by no means unusual, while in a different field altogether there were many German commercial clerks employed in the Manchester area. But there were also young Germans and Swiss who came to England with high scientific training and who were prepared to sell their talents in the great industrial areas in the north. Indeed, these young men formed such a substantial proportion of the first applied, professional scientists in England that they seem to have made the vocation almost a foreign monopoly; in much the same way as music and music teaching had, by tradition, come to be considered to be the province of Italians. The significant point was that the English were not yet producing professional scientists of their own, whereas Continental nations had been doing so for many years. To assert that this was because the English did not 'value' science is to beg the question; it certainly does not explain why, if this were so, the foreign experts bothered to come over—men like Lunge, Caro, Martius, Hurter, Griess, Auer, Obach and many others. Or, to put the point directly, had the English educational institutions provided for the training and employment of such men, would not the English industrialist have become more scientific at a much earlier date, and so have achieved the applied science revolution without foreign aid? To blame the industrialist for failure to appreciate science and research at a time when the higher educational institutions of the country did little to forward original work was both unjust and foolish. The reason for such condemnations seemed to lie in the belief that a 'demand' for 'science' should somehow arise from the heterogeneous mass of industry; this would ultimately clarify itself and result in a suitable supply of appropriately trained scientific experts. When, from the confusion of industrial voices, no such demand arose, it was felt that appreciation was lacking, and, in consequence, exhortation was called for.

THE TECHNICAL EDUCATION MOVEMENT

The Royal Commission of 1881–4 had found that nowhere in Europe was night-school instruction in science carried out on a scale comparable with that achieved in England by the Department and the City and Guilds Institute. They observed also the strictly scientific work carried out in German and Swiss chemical works and the high scientific standards of Continental works managers. It is very strange, therefore, that in the years following the publication of the Commission's reports, the technical education movement emphasised night-school science rather than what may be called the higher technology and applied science.

There was a curious vagueness surrounding the ideas of the leaders of the movement that makes it difficult to specify with any precision the objectives they had in mind or to assess the ultimate effects of their movement. In 1887 Huxley had called for further technical education as an instrument in the struggle for national existence [2]; and, in 1886, Roscoe, with the backing of Acland, had proposed the formation of the National Association for the Promotion of Technical Education (1 July, 1887) with the aim of ensuring that the recommendations of the Commission on Technical Instruction were carried out. Huxley, addressing a meeting in Manchester in the following November [3] demanded more practical science; what were needed, he said, were not professional mathematicians, chemists and physicists, but people broadly familiar with the underlying scientific principles of industry; for there is 'hardly a branch of trade or of commerce which does not depend, more or less directly, upon some department or other of physical science, which does not involve for its successful pursuit, reasoning from scientific data'. But this generalisation is so wide that it is difficult to see what value it has: what, for example, is meant by 'reasoning from scientific data'? In maintaining that professional scientists were not needed, Huxley has been proved by events to have been quite wrong. Indeed the Commission of 1884 had found that the professional applied scientist was already at work in Germany. Also, the notion of 'the sciences underlying the arts' was, by that time, archaic, and, in this respect, Huxley's views do not show any advance at all on those of George Birkbeck, Benjamin Heywood or even Adam Smith. Huxley admitted that technical education could not be easily defined but he excused himself on the grounds that such education would be experimental for years to come. It seems, in short, to have been the case that many enthusiastic advocates reduced themselves to the position of proclaiming 'We cannot say what technical education is, but we must have it'; and when it is remembered that virtually no subject of tuition* can be described as non-technical the vagueness of the demand becomes more apparent. What, in fact, was needed, as the Commissions of 1782 and 1884 had urged, was systematic secondary education, and perhaps many supporters of the movement used it as a stalking horse to secure this.

* The Classics can be considered as strictly technical: e.g. for teachers.

However the impossibility of saying in relatively few words what was meant by technical education aroused the critical faculties of two men who were certainly not reactionaries of the old school. One of these men was an Oxford classicist and Professor of Humanity at Edinburgh [4]. Examining the findings of the Technical Commission, Professor G. G. Ramsay concluded that where Britain was decisively backward was in the development of scientific chemistry as applied to coal-tar colours and to agriculture, especially the sugar-beet industry. This, concluded the classicist, was due to the possession by Germany of a corps of professional scientists. This seemed undeniable for every economic advantage was on the British side to begin with; she possessed the raw material and could produce it more cheaply than Germany (indeed we exported to Germany the coal-tar that they used to make the aniline colours), she had the capital, the transport facilities, in fact, everything bar the chemists. It is regrettable, but true that in this instance the classicist rather than the biologist seems to have understood more clearly the importance of true applied science. In any case a powerful ally appeared on Ramsay's behalf. This was Lord Armstrong (W. G. Armstrong) who stressed the desirability of public science laboratories, but denied the need for widespread technical education [5]. To this Playfair replied a month later, and Magnus in the following November. It must be admitted that Playfair's case was not convincing [6] and, to Armstrong's rejoinder [7] he made no reply.

However diffuse the objectives of the leaders might have been, they were certainly able to make piecemeal progress. The National Association criticised very effectively the payment-by-results system, quoting the case of the 16-year-old boy at Bradford Technical College, who, in May 1888, passed in no fewer than 19 subjects in the Department's science examinations, while at least three had passed in 18 subjects and many in 16 or 17. This, they concluded mildly, was hardly 'wholesome'. Supporters of the movement like H. E. Armstrong and Philip Magnus, disciples (in effect) of Edward Frankland, worked hard to develop a new education; stamping out 'test-tubing' in chemistry classes and calling for effective inspection in place of payment-by-results.

A development parallel to the technical education movement and one that to a great extent, later merged with it, was the Polytechnic movement. The latter originated with Quinton Hogg's educational charities, especially with the institution he acquired and rejuvenated in 1881: the Regent Street Polytechnic. In 1883 James Bryce was able to get the City Parochial Charities Act passed, whereby various charitable endowments could be used to found additional polytechnics on the Regent Street model. The purpose of these new institutions was defined as: 'A corporation of working men and women, bound together by the sympathy of kindred occupations and bent on mutual improvement by means of such agencies, other than religious, as are calculated to promote their intellectual, physical, moral and material well-being'. By 1898 eleven of them had been founded in London alone, under the administration of the Charity Commissioners, representatives of the Guilds and the Technical Education Board of the L.C.C.

In 1889 the first Technical Instruction Act was passed and for this Roscoe, who had entered parliament as a Manchester radical in 1885, deserved much credit. This act empowered county and borough councils to levy rates to establish technical schools for teaching 'the principles of science and art applicable to industries' and the 'application of special branches of science and art to specific industries and employments'. In the following year Goschen's Local Taxation Act placed at the disposal of the local authorities a sum of £750,000 specifically for technical education. The battle for State-aided technical education had been won, and the scene was now quite modern with polytechnics, technical colleges, etc. The mechanics' institutes were no longer a living issue.

A powerful influence in the public acceptance of technical education was the increasing awareness of the development of foreign competition. With this threat very much in mind, the National Association, in 1889, circularised manufacturers, asking for the views of 'practical men'—much as the Society of Arts had done, nearly forty years before. To their inquiry Ivan Levinstein, of Manchester, one of the few still engaged in coal-tar colour products in England, replied that what was needed were better trained chemists: quality before quantity; for many of our technical school chemists were badly trained. On the same occasion Sir William Mather, of Mather and Platt, expressed the very liberal view that 'In the technical schools the means should be provided to pursue science in the abstract for original research with attendant general culture'. Mather had a right to be heard for not only was his firm one of the largest engineering concerns in England, but they had a splendid record in education; they had long ago established their own trade school and had compelled their apprentices to attend classes for two hours every day. Roscoe and Acland agreed with Mather's recommendation and suggested that the solution was government grant aid for the university colleges.

THE UNIVERSITY COLLEGES

Apart from Scottish and Irish colleges, the first university college to benefit from government aid was the new University College of Aberystwyth. But the English university colleges, too, were not in a very healthy condition; in April, 1887, Jowett drew attention to the fact that the endowments were insufficient for the full development of the colleges; and the lower middle classes could not afford a university education.* On 10 March, 1888, a delegation that included Playfair, Roscoe and Acland waited on Lord Cranbrook with the request for a grant of £4,000 a year for each university college. The government replied with an offer of £15,000 a year for all; an allowance equal to the annual State subsidy of the Zürich Polytechnic. The award was to last some five years, after which it would be reviewed and a prototype University Grants Committee was set up to administer the money.

With such advanced sentiments in the air it is not surprising that there was growing discontent with London University; that emasculated examining board created by the utilitarian theorists of 1837–58. Matthew Arnold had called for reform, as we have seen; Lyon Playfair not long afterwards, and there were also the numerous critics of the expanded examination system. These critics clarified the issue when in 1884, they founded an organisation to promote the establishment of a teaching university in the capital.†

A major reform was made inevitable when the two chief colleges, University and King's, petitioned the Queen for joint university status, the new University to be called the 'Albert University'. This proposal was seen to be sufficiently important to justify the appointment of a Royal Commission with Lord Selbourne as Chairman and five other members, three of whom were distinguished scientists: G. G. Stokes, J. E. C. Weldon and Kelvin. Unfortunately, this first Commission proved a most unsatisfactory one, for it rejected the petition of the colleges while recommending the reform of the University by the inclusion of college representatives on the Senate [δ]; a decision from which the three scientists rightly dissented; adding a rider to the effect that they would have preferred a teaching university.

* Report of conference of university delegates.
† This had the backing of Mrs. E. G. Anderson, Grylls Adams, James Bryce, Warren de la Rue, Dewar, Carey Foster, Frankland, Huxley, Henrici, Lankester, Lister, Lockyer, Playfair, Kelvin, Tyndall, etc., as well as the Heads of many Oxford and Cambridge Colleges.

However, we are not concerned with the legal disputes that accompanied this investigation but with the development of scientific institutions in London. An analysis of the figures for the London B.Sc.(Hons.) awarded between 1880 and 1900 inclusive is illuminating and demonstrates the force of the claim made by the two colleges:

University College	124	Oxford and Cambridge	56
King's College	14	Provincial Colleges	219
Other London Colleges	127		
Totals	265		275

While the University clearly did not honour its parents, these had become almost minor universities in their own rights; certainly in terms of student numbers and work—i.e. research—carried out. But they were too poor to be able to endow research laboratories on the scale achieved by Oxford and Cambridge universities,* and the London University, as a purely examining body, obviously could not do so; the latter had been founded at a time when the only overt function of a university, as opposed to colleges, was to examine. (Lowe and those who thought like him, had maintained that the only duty of the State was to supervise the examinations, and even Pattison, with no faith in examinations, had believed that 'it was not the duty of the State to subsidise science through the universities'.)† Yet, in spite of neglect and poverty, the London colleges achieved a remarkable record in the annals of nineteenth-century science.

Karl Pearson had already pointed out that the London degree system amounted in practice to a complete denial of the principle of *Lehrfreiheit*. That is, the teacher could not, if his students were to be successful, depart from the strict course laid down by the remote examination board over which he had no control. Lister and Lankester had informed the Selbourne Commission that only a minority of London students took the university degrees because they preferred to follow the curricula of their teachers rather than those of an outside body. Students who took the university degree naturally concentrated on the requirements of the examiners rather than the ideas of the teachers.

* Both universities found the creation of research laboratories very expensive.

† *Academical Organisation.*

Educational deficiencies of this kind, in the capital city of the wealthiest country in the world, could hardly be tolerated. Accordingly the next Royal Commission, the Cowper Commission of 1894 [9], stressed in its report the need for scientific research institutions in London '. . . for those advanced students who cross the Channel to find what should be available here'. A reformed University of London, they urged, should have adequate State support. Large sums of money are spent on science in Germany; at present, London is behind Zürich. The unfortunate history of the aniline dye industry did not escape the Commissioners. Like so many others before them, they found that the facilities for training the research chemist were far better in Germany than they were in England. (It was, by then, a *sine qua non* for British chemists to append the German Ph.D. to their names.) As for physics, as Olaus Henrici told them, the higher teaching and advanced courses comparable to those given at German universities by authorities like Quincke did not exist in London at that date.

A local comparison was made by Sir William Ramsay. Justifying the claim of University College to be considered as a university, he showed that the volume of research at the college equalled or exceeded that done at Oxford or Cambridge; for, between 1890 and 1892 no fewer than eighty-four scientific memoirs were published. Ramsay had a splendid scorn for book-learning; the London B.Sc. was harmful for research; the emphasis was wrong and it was not practical enough. Research training should begin at student level. As for examinations, Ramsay believed that the ability to pass them, to acquire material and then reproduce it when required, might be a good qualification for a barrister or a government official, but not for a scientist. He needed developed inventive powers.

For Huxley, now the venerable elder statesman of science, university education was still to be liberal; degrees were of no importance for men of science—only for professionals; in fact, Huxley knew of no leading scientist who cared whether he had a degree or not. An interesting point of view that would seem to indicate that Huxley's mind must, at that time, have been fixed more on the past than on the present; and it also suggests that he had very little idea of the trend of future developments. However, he concurred that our laboratories were nowhere up to German standards; a point also made by Hudson Beare who emphasised the importance of the Massachusetts Institute of Technology. Ambrose Fleming observed the development of German industrial laboratories, as did Magnus who gave Playfair as his authority for the statement that German industrial laboratories were now employing up to fifty chemists on research. With this was correlated the sugar beet and aniline dye industries (cf. G. G. Ramsay). In spite of this it was difficult to get adequate State support for the laboratories of the Royal College of Science. 'They are', remarked Roscoe, 'a national disgrace.'

The German industrial laboratory was further discussed by H. E. Armstrong and the inevitable comparison drawn with this country. Here, thought Armstrong, the maximum number of chemists employed in any one establishment was about six (he was quite right). Only brewing appeared to be progressive, as the work of Griess and Brown showed. The real beginning of German development was the Giessen laboratory and, later, the work of Hofmann when he returned from England (see p. 101 above).

Apart from Huxley, the last ditch defence of the liberal education was left, appropriately enough, to two Irishmen: the theoretical physicist G. J. Stoney and the Rev. Dr. Mahaffy, a classicist. Stoney, a university administrator of great experience* as well as an able physicist, condemned the common practice of training men to a very high standard in particular, or specialised, departments of knowledge. He approved Archbishop Whateley's epigram: 'it makes men as narrow as they are deep'. It is a bad thing that thought should be guided by narrow men for it must result in a disintegration of studies; it should be the duty of a great University of London to counteract this disastrous trend. Present university arrangements direct far too much attention to examinations, with correlative specialisation. For his part Dr. Mahaffy conceded that specialisation was unavoidable but he urged that more than one subject should be studied. From personal experience he, too, believed that university studies were too specialised. It was, he thought, regrettable that special studies now began with the student's undergraduate career, for there were too many specialists. The increase of knowledge does not necessitate specialisation; it is characteristic of great men that they are widely cultured.

In the meantime Lyon Playfair had alleged that the backwardness of university education and the absence of scientific facilities in London meant that the best intellects either did not go there or were soon tempted away to more advanced centres [10]. Karl Pearson had said much the same thing when he lamented that there was no physical laboratory in London worthy of the capital (which was not surprising

* B. B. Kelham, op. cit.

in view of the poverty of the two main colleges). Besides the two leading institutions there were only the small government college in South Kensington, the Bedford Ladies' College and a few minor academic foundations together with the polytechnics. The remote London University examination board could not possibly give a proper incentive to advanced study and it could give even less sustained research effort. But, obvious though the deficiencies were, the old London University could still command the loyalty of a large number of men and women; it had always been 'open'; no one had ever been excluded on grounds of faith or class; it had been the first to open its doors to women; it had pioneered the academic study of English literature; it had been the first to introduce diplomas in teaching and it had awarded the first science degrees.

However, the movement for the reform of London University was now so strong—it enjoyed the support of so many individuals as well as of learned societies—that it could hardly be ignored. The truth was that the utilitarian examination-university, notable as its services had been, could not meet the educational and scientific requirements of the times. Accordingly, an Act of 1898 based on the reports of the Cowper Commission, instituted a Statutory Commission charged with the task of implementing the proposals for reform.* Ultimately the university was reconstituted much as it is to-day: the internal division was created and the vestigial remnant of the work of the Utilitarians remains in the form of the External degree.

At the end of the first five year period the question of the universities grant came up again. A Treasury minute of the 5 July, 1894, resolved that the grant was to continue but, on the 28 December, 1895, an influential deputation waited upon the Chancellor of the Exchequer and requested not merely renewal but an increased grant as well. Thereupon the President of Magdalen College, Oxford, and G. D. Liveing (Cambridge) were asked to inspect the colleges and report on the state of the studies pursued therein. Their report was wholly favourable; but they pointed out that the endowments were too small, and, in some cases, the staff too few; although staff appointments were providing young men with valuable opportunities for research and learning. Scholarships were neither liberal nor many and the regular students were middle class; although the training college

* Very shortly after this, the federal Victoria University disintegrated and the constituent colleges of Manchester, Liverpool and Leeds became independent universities.

students were usually working class. At the same time a parallel enquiry by Robert Chalmers, a Treasury official, showed that the only university college to return a surplus was Bedford College (£613); the total overall deficit amounted to some £10,000. Acting on these recommendations and findings the government increased the Treasury grant to £25,000 a year.*

Four years before the Cowper Commission sat, two administrative steps had been taken in 1890 which were to be of considerable importance for the future of the university colleges. Elementary school science, which had been urged for many years by reformers like Huxley, was in that year incorporated in elementary education by a change in the educational code and provision was also made for manual training. In the same year, the day training colleges† that were attached to some university colleges were granted official recognition. As a consequence the number of training college students doubled and similar training colleges were established, in 1891, at Cambridge University, Leeds, Liverpool and Sheffield and, in 1892, at Oxford University, Bristol, University College (London) and Aberystwyth. By the end of the century they had over 2,000 students and many of these were reading for university degrees. Correspondingly the number of science departments in elementary schools increased from 173 in 1891 to 1,396 in 1895 [11].

While this was in progress, the agitation against examinations was continuing. An appeal was launched in 1888 against the 'Sacrifices of Education to Examinations'‡ [12]. According to Max Müller, 'It is the best men who suffer most from the system of perpetual examination. The lazy majority has been benefited but the vigour of the really clever and ambitious boys has been systematically deadened'. It was in the natural sciences, above all, that examinations multiply most, thought Professor E. A. Freeman: they proliferate, split up or 'specialise' and so a new '-ology' is born. With even more pessimism Frederick Harrison doubted whether any great historian, scientist or lawyer could have got a First in modern examinations, for he would lack the 'smart, cocksure style of the trained examinee'. By February

* On 25 April, 1897.

† Such colleges were attached to King's (London), Mason, Durham, Owens', Nottingham and Cardiff.

‡ The signatories including Samuel Alexander, Lord Armstrong, James Bryce, R. B. Clifton, Sir Edward Grey, R. B. Haldane, Halford Mackinder, Karl Pearson, General Pitt-Rivers, F. J. Romanes and E. B. Taylor as well as many other public and university men.

1888 over a hundred M.P.s had signed the protest. Sir Frederick Pollock, himself an examiner of some experience, could say, '. . . we have, indirectly, discouraged every kind of intellectual activity which has not an obvious bearing on [examinations] . . . "Will this pay in the Schools?" is the inevitable check on both learners and teachers'.

The most determined critics of all were H. E. Armstrong and Norman Lockyer: at least from the scientific point of view. Armstrong was an enthusiastic advocate of the 'heuristic' system of teaching chemistry* and as he had done so much to develop higher technological education he may, perhaps, be forgiven for describing examinations as 'The execrable system we have allowed to grow up . . .' [13]. The only examination Norman Lockyer had ever passed had been one for a War Office clerkship. He looked upon examinations as 'the poison of education' [14], and this was an opinion from which he never deviated. Lockyer had very liberal views; he saw no reason why university education should not be free, thereby ending the baneful scholarship examination system and he looked forward to the future university as resembling the medieval model; open and patronised by wandering scholars.

GERMAN COMPARISONS

For a wide variety of reasons German universities were not strictly comparable with English ones. Accordingly when Bryce asserted that in 1882–3 Germany had 24,000 students to England's 5,500, he limited English students to those attending Oxford, Cambridge, Durham and Victoria; and did not count those in the London colleges, etc. But whether or not comparisons could accurately be made, there can be no doubt that the German universities did, over the middle of the century, undergo a great expansion; at the same time, the philosophy (science and art) faculty increased within the university. Thus in 1830–1 the percentage of philosophy students was 17.7 while in 1881–2

---

* The heuristic system was based on the principle that the pupil should learn chemistry by the methods of discovery for himself rather than by memorising facts. Experiments were designed to ensure that the pupil could experience, to some extent at least, the moments of private illumination that come to all engaged on scientific research. The difficulties of the heuristic system were that it was expensive, time-consuming and demanded a high degree of skill on the part of the teacher. It was also very difficult to compare and mark the relative performances of different pupils.

I am grateful to Dr. W. H. Brock and Mr. S. Gill for some profitable discussions about the heuristic system.

it was 40.3.* This was largely due to the development of science for, while in 1841 13.6 per cent of the philosophy students did science, in 1881 the percentage had risen to 37.1. This increase was partly, but not wholly, to be ascribed to the demand for trained teachers of science caused by the development of instruction in the various grades of schools. In round figures the number of science students had risen from 290 in 1836–45 to 3,000 in 1881–4; more than the total of science graduates in England.

There occurred, therefore, a rapid increase in the number of German students able to enjoy a scientific education superior to that generally obtainable in England. That this would have an appropriate effect on German industry was the inference that caused four veterans of the Technical Education Commission of 1884—Magnus, Redgrave, Woodall and Swire Smith—to visit Germany in 1897. Their subsequent report was considered sufficiently important to be published as a Blue-Book [15]. In 1884 they had been inclined to doubt the value of the polytechnics;† now they were fully converted. What especially aroused their admiration was the magnificent new physics institute at Charlottenburg: '. . . probably the most complete institute in Europe for physical research'. They paid proper tribute to the aniline dye industry but they were also very impressed by the development of the electrical industry and the related development of the electro-technology departments in the polytechnics. In 1884 there had been no electrical laboratories in Germany to rival those at the Finsbury College; now England had nothing to compare with the laboratories at Darmstadt and Stuttgart. Although the technical education facilities for the German worker were still inferior to those available in England, the German primary and secondary educational machinery was much superior.‡

* J. Conrad, *German universities over the last fifty years* (London, 1885).

† It was said that, prior to 1871, the German States had vied with one another in founding polytechnics. Consequently, upon the Bismarckian unification, it was found that there were too many. While there may have been some truth in this for a few years, Germany soon found that she needed all her polytechnics and new ones as well.

‡ It is interesting to recall that Sir John Clapham maintained that the rapid development of the German electrical industry from 1883 onwards was, beyond question, the greatest single industrial achievement of modern Germany. They led the way in solving the problems of electrotechnology and the specialised applications of electricity. Before the 1914 war Germany was selling electric cable, etc., to England.

From the point of view of industrial science the most remarkable German achievement was still the aniline dyes industry, for it was the very first of the truly scientific industries; its precedence as such over the new electrical industry amounted to some twenty years. For this reason, at the end of the century, Dr. Frederic Rose, British Consul at Stuttgart was authorised to investigate how 'German chemical industries have benefited from the sums expended by the German states on chemical instruction' [16].

Taking Prussia alone whose population was slightly less than that of England and Wales, Rose found that the university philosophy faculties had increased from 1,155 students in 1850 to 5,528 in 1899 while, in the latter year the Prussian government granted the eight universities a subsidy more than ten times that granted to the English university colleges. In 1899 Germany had 33,000 fully matriculated university, and 11,000 polytechnic students; the latter population having doubled in seventeen years. With this correlated the fact that two-thirds of the world's annual output of original chemical research came from Germany.

Rose was informed that Germany had 4,000 trained (university or polytechnic) chemists in industry; the largest single employing industry being organic chemicals (aniline dyes, etc.) with 1,000 chemists.* Carrying out a sample investigation, he found that 67 works employed 1 chemist, 33 between 6 and 20, and 9 dye works between 20 and 105. In 25 years the number of chemists had more than doubled. Of 633 chemists employed in 83 concerns, he found that no fewer than 436 were university Ph.D. graduates.

These figures give ample proof that applied science, properly speaking and as defined above (pp. 15–16) had been achieved in Germany long before the end of the nineteenth century. It had all begun in 1862 with the founding of the aniline dye industry. As early as 1868, Meister, Lucius & Co. and the Badische Anilin Co. were employing university chemists but their precise functions are unknown, for with the exception of F. Bayer & Co. the records of the pioneer German firms are inadequate [17]. There were good reasons why industrial research laboratories should have developed early on in this science-based industry in Germany. For one thing, the synthesis of alizarine and the limitless possibilities opened up by the azo dyes put a premium on

---

* Other industries were: Agricultural stations and Laboratories, 700; Inorganic, 250; Sugar Refining, 300; Smelting, 400; Artificial Manures, 90; Various, 600; etc., to a total of 4,000. (1897: Prof. Fischer.)

effective research; again, as J. J. Beer has pointed out [18], the dye-stuffs industry was subject to fluctuations due to the 'appetite for novelty characteristic of the fashionable world. New colours enjoyed only brief popularity.' And then, the political unification of Germany resulted in a unified set of patent laws for the whole empire; this made the copying, previously practised, impossible so that firms had to pioneer if they wanted to stay in business [19].

The records show that F. Bayer & Co. did not find it particularly easy to develop an effective research laboratory. But after initial set-backs they found the right man for the job of research director and their laboratory became efficient and profitable. In 1881 they employed some 15 chemists; by 1896 the number had increased to 104; at both dates probably about one-fifth of these men were employed on research. This remarkable growth was made possible by the building of a new research laboratory for which $1\frac{1}{2}$ million marks had been allocated in 1890. The new laboratory possessed a library and had an auxiliary staff of glass-blowers, analysts, technicians and lab. boys whose function was to relieve the research chemists of tedious and time-wasting work [20]. Evidently the establishment of industrial research laboratories depends on a complex variety of technical, social, political, legal, scientific and commercial factors.

Underpinning, and therefore making possible, this advanced indus-trial technique was the great educational system of Germany; and, not least, the original Giessen laboratory and the vision of Wilhelm von Humboldt. Science was now being directly and systematically applied to industrial problems exactly as Playfair had envisaged at the time of the Exhibition. There was, then, some substance in Ostwald's boast to Sir William Ramsay: 'The main point of our system may be expressed in one word—freedom—freedom of teaching and freedom of learning. . . . As for the inventive man of original ideas, it has often been proved that for him any way is almost as good as any other for he is sure to do his best anywhere.' Ostwald described how the student performs his original research: '. . . he generally learns much more than he has heard at lectures. Every part of the investigation forces him to revise the scientific foundations of the operations he performs.' With considerable justice, Ostwald concluded: 'The organisation of the power of invention in manufactures and on a large scale is, so far as I know, unique in the world's history, and it is the very marrow of our splendid development'.

Measured by the scale of national industry the number of German chemists is, no doubt, small. But their importance lay in the future; the number of industrial chemists is now far from small and their national importance is infinitely greater. The credit, therefore, belonging to German chemical industry was that of pioneering applied science on a notable scale. The subsequent development of the German electrical industry is, in this sense, less outstanding. They already had their universities, polytechnics and trade schools with the correlative professional scientists and they had the example of the chemical industry—what could be achieved—immediately before them. In fact, it would have been remarkable had they not been able to develop a highly scientific electrical industry.

The achievement was, as Ostwald claimed, substantially unique. Only the French efforts at the time of the revolutionary wars can be said, in some measure, to have anticipated the German development of applied science; but the former was artificially stimulated and once the central initiative lapsed the effort subsided; the latter was a natural development, not depending on inspired government but on the organisation of society. As such it marked a revolutionary stage in the development of industry, and once the uniqueness of the event was recognised, German pride and foreign envy became the order of the day. Sometimes these emotions were carried a little too far; thus Michael Sadler could write, in 1899, 'The organisation of modern life in Prussia has been dominated by scientific conceptions; not, that is, by any exclusive regard for physical science in its narrower sense, but by those ideas of exact and co-operative enquiry and endeavour which have been so brilliantly illustrated, and therefore so powerfully enforced, by the advance of modern science'. The initial cause of German scientific greatness was not, according to Sadler, efficient organisation, 'but an intense and self-sacrificing enthusiasm for truth'.

The Prussian way of life, scientific or otherwise, has passed into history. But the second of Sadler's generalisations is more plausible; we may compare it with the words used by A. W. Williamson in his presidential address to the British Association of 1874: the fact that so many English scientists were then of independent means 'proves', thought Williamson, 'how true and pure a love of science exists in this country and how Englishmen will cultivate it when it is in their power to do so'. One must agree with Williamson; a great and self-sacrificing love for truth was repeatedly demonstrated by British scientists: witness Faraday, Joule, Dalton, Darwin and many others.

I am not concerned to make generalisations as to national character; I am only concerned to show that any such may be misleading when applied to science and those who cultivate it. The truth would seem to be that Germany built up an organisation that was liberal enough to accommodate and intelligent enough to foster the activities of truth-loving and knowledge-seeking Germans. The truth-loving Englishman, on the other hand, had to make his own way in the world and had no shielding or fostering organisation.

APPLIED SCIENCE IN ENGLAND

The merits of a true and pure love of science had, however, proved insufficient to enable important branches of science to flourish in face of the increasingly brisk competition of the professional Continental scientists. In 1884 W. H. Perkin, in the course of his Presidential address to the Chemical Society, commented on a disturbing trend [21]:

'The first thing that attracts attention is the startling and anomalous fact that the number of papers read before the Society (and I think this may be taken as a good criterion, especially as but few have been brought before the Royal Society) is declining year by year. The largest number we ever had was in the session 1880–1 when there were 113 communications brought before us but in 1881–2 they had declined to 87; in 1882–3 to 70 and this last session to the lamentably low number of 67, or about the number we had nine years ago.'

The country that, in the first two decades of the nineteenth century had boasted of Dalton, Davy, Prout and Wollaston as its leading chemists could by this time point to no one of major stature. Playfair and Williamson had long been lost to administration, Roscoe had turned to politics and Frankland was well past his prime. No new talent of any reasonable promise had appeared and measured even by the crude yardstick of numbers failure was apparent.

The decline in academic chemistry was accompanied by a decline in the advanced sectors of chemical industry. The small, undercapitalised firms in the synthetic dyestuffs industry in Britain were unable to compete with the new German firms despite the obvious advantages that they enjoyed, not the least of which were plentiful and cheap basic materials supplied by a well developed heavy chemical industry and close proximity to an enormous home market [22]. Between 1869 and 1873 Perkin's firm had expanded very considerably in response to the demand for synthetic alizarine. But when he sold out the following year competition had greatly increased, prices had fallen and there was an urgent need to expand, diversify and in fact reorganise the firm. The essential steps were not taken and although Brooke, Simpson and Spiller were able to recruit two leading colour chemists—Raphael Meldola who was with them from 1877 to 1885 and A. G. Green who stayed from 1885 to 1894—both men felt that their talents were being wasted and resigned [23]. The firm itself went bankrupt before the end of the century. By then both Roberts, Dale and Williams, Thomas and Dower had disappeared so that the only British firms left were the British Alizarine Company (which had taken over Perkin's old plant), Read, Holliday, the Clayton Aniline Company and Ivan Levinstein's. The last, in spite of being owned and run by a German, did little research and was reputed to have been a bad employer of scientific labour.

The reasons given for Perkin's retirement were his own desire to return to academic-type research and, according to two of his sons, the near impossibility of recruiting organic chemists to carry out research on new dyestuffs. German chemists could be recruited but they tended to return to Germany after a while and thus were lost to the competitors.

This explanation, however, rings only partly true. Even if German chemists were rarely willing to stay for long periods there was no reason why Perkin should not have trained his own research workers.* The chance to work under the leading British organic chemist who had also founded a unique industry should have attracted an adequate number of able young men. In a situation that was analogous but in circumstances that were much less favourable James Watt and Matthew Boulton had recruited able young men from all over the country, so that their Birmingham works became known as the 'science school of Soho' [24]. Finally, it is clear that while Perkin at one time employed four chemists in his works he always reserved the research function for himself and the research laboratory was kept strictly private.†

The conclusion must be that Perkin retired partly because he wanted, as he said, to devote himself solely to research and partly because the managerial requirements for dealing with an enterprise of growing complexity and technological refinement were beyond his resources.

* There was a precedent for this. R. C. Clapham told the Parliamentary Committee on Scientific Instruction (London, 1868) that Tyneside alkali works commonly took on apprentices for training in their laboratories as (presumably) control chemists. The facilities at Durham University were said to have been inadequate.

† Alfred Nobel was, in a similar way, the main source of ideas and new researches for his dynamite works.

The pressures of innovation were great, competition was mounting and the synthesis of alizarine meant that one of the two main traditional dyestuffs (madder) had fallen to the new industry; the synthesis of the other (indigo) could, perhaps, be foreseen. There is no point in finding new products unless you can manufacture them economically and sell them at a profit. To meet all these requirements Perkin would have needed not only research chemists but also technical managers, engineers, chemical engineers and properly trained technical salesmen. Where could such men be found in England at that time?

Some light is cast on this by the experiences of the suppliers of specialist labour, the university and technical colleges. The Yorkshire College of Science was, as we remarked, founded in 1874. Four years later, following a grant of £15,000 from the Clothworkers' Company, the College set up a Dyeing Department and appointed J. J. Hummel, a graduate of the Zürich Polytechnic, to be the first Instructor and, later, Professor. Shortly afterwards, in 1883, the new Bradford Technical College (now Bradford University) put Edmund Knecht, also a graduate of Zürich Polytechnic,* in charge of the dyeing courses there.

In the first year of Hummel's Department material donations were made to it by the Badische Anilin Company as well as by a few English commercial firms. In the following year 'B.A.S.F.' were joined, as contributors, by Meister, Lucius & Brüning and by Bayer & Company. In 1882 Leopold Cassella followed suit and thereafter the four big German firms were regular donors. Read, Holliday of Huddersfield made their first contribution in 1883 but they were never as regular as the German firms. Very shortly after this other names, which are also familiar, began to appear in the lists of donors: Geigy, Agfa, Ciba, etc. Finally, in 1898, Levinstein's began to contribute; late in the day, but then, of course, they were on the far side of the Pennines.

Although Hummel was evidently a capable chemist and although Perkin's second son, A. G. Perkin, soon joined the staff, the Department was not very successful during the first twenty years of its

* E. Knecht (1861–1925), the son of Gustav Knecht of Liverpool, was educated at Zürich University and Zürich Polytechnic where he took his Ph.D. in 1882. In 1890 he moved to Manchester where he became Professor of Technological Chemistry at Manchester College of Technology (now U.M.I.S.T.). In collaboration with J. B. Fothergill he wrote a standard work on textile printing (1912).

J. B. Fothergill's background war mainly practical; he had worked for Frederick Steiner (Accrington) and S. Schwallu (Middleton).

existence. The reasons for this were variously said to have been the resistance to innovation offered by the all-powerful dyeworks foremen, the growing competition from similar departments as they were established in other colleges and the inadequate scientific training of the students entering the Department. Another and revealing reason was given in the course of a confidential report that Raphael Meldola drew up in 1902 for the Clerk to the Clothworker's Company, Sir Owen Roberts [25]:

'The manufacturers of colours are themselves—especially the Germans—not only keeping the dyers supplied with a constant succession of new colouring matters, but it is the custom of the German firms to send round their own experts to teach the dyers how to use these new products or to issue such detailed instructions that practically nothing is required of the dyer in the way of scientific knowledge.'

This tends to confirm the supposition that German firms were able to make effective use not only of research chemists, without whom they could not have maintained their stream of innovations on which supremacy rested, but also of a body of highly trained technical salesmen. In these circumstances it was hardly surprising that many of the best products of the Leeds school went either to the German firms or to be teachers in the technical colleges which, following the Act of 1889, were then in process of expansion. One who did not take either of these courses was a Mr. C. M. Whittaker who, in 1899, joined Read, Holliday as an assistant chemist and later rose to be head of the research laboratories [26].

Read, Holliday had, for some time, been recruiting such scientists as they could find—the majority appear to have been German—and by 1890 they had a research laboratory in which about four or five chemists were continuously employed on research. This *may* have been the first industrial research laboratory in Britain. Some caution is, however, necessary here. The industrial record is notably incomplete for it is common practice for firms to destroy their archives. They may do this because a new office block has been built, or because they have been taken over by a larger concern, or because they want to make more efficient use of the space available. Indeed there are a multitude of good reasons why this sort of thing should happen: after all, firms are not in business for the benefit of historians and archivists. Moreover, even where records are available it should always be remembered that what one man may describe as a 'research laboratory' another may,

with apparently equal authority, call a 'development department'.*
Who is to say, when we are dealing with organisations of eighty or
more years ago, which description is correct?

If, then, we tentatively concede that Read, Holliday of Huddersfield
had the first industrial research laboratory, the second was almost
certainly at Widnes in Lancashire. In November, 1890, a rationalisation
of the heavy chemical industry resulted in the creation of the United
Alkali Company, an organisation which included 48 firms in Scotland,
on the Tyne and in Lancashire and whose Chairman was Sir Charles
Tennant. The Chief Chemist of the new group was Ferdinand Hurter,
who had previously been Chemist to the Gaskell-Deacon Company.
Hurter, a Swiss and a product of Heidelberg University and the Zürich
Polytechnic, had contributed a great deal to the development of the
Deacon chlorine process. He was a true scientist and Dr. Hardie, the
historian of the chemical industry at Widnes, tells us that, prior to
1892, 'Thermodynamics were particularly useful in Hurter's hands for
the calculation of heat wastages in the various Leblanc processes and
in the determination of the optimum temperatures for the reactions
they involved. It must be remembered that what is to-day routine
procedure was, in Hurter's time, almost unknown: his application of
the principles of "pure" science to the processes of manufacture was,
in great measure, the work of a pioneer.'†

One of the first acts of the new group was to investigate the possibilities of establishing a central research and analytical laboratory. The
plans Hurter drew up were approved and, by 1892, the laboratory was
at work. The original staff numbered 'half a dozen chemists, a general
handyman and a confidential clerk'.‡ The first applicant for appointment under Hurter was H. Auer, also of Widnes and also a product of
the Zürich Polytechnic.

A contemporary comment made by *The Times* and quoted by Dr.
Hardie was to the effect that 'the chemical industry owes nothing to
the historic educational institutions of this country'—but something
to institutions like Zürich Polytechnic and the German universities.
Indeed, the influence of the latter can be seen in the establishment of a
third early industrial research laboratory: at the Nobel explosives

* This observation is based on personal experience of modern, science‑
based firms.

† D. W. F. Hardie, *A History of the Chemical Industry at Widnes* (I.C.I., The
Kynoch Press, 1950), p. 169.

‡ Ibid, p. 176.

factory at Ardeer in Scotland. Here, from about 1894 a small group of chemists was engaged on systematic research under the general supervision of Joseph Sayer, who had been trained by Dittmar at the Andersonian Institution (now the University of Strathclyde).* W. Dittmar had, in his time, been a student under Kekulé at Bonn.

With regard to physics and the related sciences the rate of progress was slower. In one sense of the word a physics laboratory had existed for a very long time in the form of the Royal Observatory at Greenwich and there was also the Kew Observatory, maintained by the British Association but endowed in the first place by J. P. Gassiot, a wealthy amateur and a President of the Royal Society. But these were not industrial laboratories. The first hint of a physics laboratory analogous to the chemical ones at Huddersfield, Widnes and Ardeer came from two electrical firms who referred secretively to their 'laboratories'. But on further examination these turned out to be one-man laboratories whose function was to test instruments.

The desirability of establishing State physical and chemical laboratories had, we remember, been urged by Colonel Strange. Kelvin had taken it up at the British Association meeting at Glasgow in 1871, alluding to the fatal neglect of science by the British government and to the disgrace of allowing such an important national activity to be the responsibility of 'self-sacrificing amateurs'. Although the Devonshire Commission had endorsed these views the next step was not taken until 1891 when Oliver Lodge, speaking at the British Association meeting at Cardiff, managed to arouse the interest of Douglas Galton and others in the proposal.

In the meantime, the Germans had, in 1895, opened the Physikalisch-Technische Reichsanstalt at Charlottenburg; the Berlin suburb in which was also located the famous polytechnic. The Reichsanstalt owed its foundation to the foresight of Helmholtz and was built on land given by the industrialist Werner von Siemens; although it was, of course, a State institution. The first President was Professor Kohlrausch, and among the early workers at the institution were Lummer, Pringsheim and Brodhun, who are known for their contributions to photometry. The total staff was between seventy and eighty.

The reaction of British scientists like Galton and Lodge was immediate; they enlisted the aid of the politically influential and scientifically eminent Lord Rayleigh [27], and a deputation was formed to

* F. D. Miles, *A History of Research in the Nobel Division of I.C.I.* (I.C.I., London 1955).

O.S.E.—7

wait on Lord Salisbury, who, as a man of science, could be expected to be sympathetic. This led to a Treasury Minute of 3 August, 1897, authorising a committee, under the chairmanship of Lord Rayleigh, to consider the proposal for a National Physical Laboratory [28]. The result was, perhaps, a foregone conclusion; the recommendation was for the establishment of a laboratory to maintain standards of national importance and for the testing of scientific instruments. But Oliver Lodge saw a little further than this: 'What I advocate has something more to do with advancing work in new directions than with stereo-typing existing practice'.

The development of the new laboratory was somewhat restricted by shortage of money, for it was, at first, more like an educational foundation than a State laboratory. It was not incongruous therefore that, in 1909, Sir Julius Wernher gave £10,000 to enable extensions to be made to the metallurgical laboratory; and Alfred Yarrow donated £20,000 to endow the famous test-tank.

Bearing upon these applied science developments of the last years of the century was Lyon Playfair's last service to science. As Honorary Secretary to the Royal Commission of 1851 (during the years 1883–9) he had been responsible for converting an annual deficit of £2,000 into a surplus of £5,000. The problem now was what to do with this money. To decide this question Roscoe, Mundella, Lockyer, Huxley and William Garnett were elected to a small committee with Playfair as chairman. At Playfair's request Lockyer drew up a scheme of scholar-ships, the essential condition for which was that they were to be awarded for research and not for 'instruction'. He had at least one good example before him, for in 1881 Manchester University had instituted the Bishop Berkeley Fellowships for research. In the 15 years during which these were available some 43 Fellowships were awarded; 29 in science, 10 in arts and 4 in engineering [29].

Whether or not he was influenced by the achievements of Owens' College—which by then had by far the biggest postgraduate science school in the country—Lockyer's scheme, adopted in June 1890, was the start of the famous Exhibition Fellowships which served as models for subsequent research fellowships in all fields of learning. The Exhibition Fellowships were almost immediately successful; among the first to hold one was young Ernest Rutherford, and since then many other leading scientists started their careers in this way. The interest attaching to the small committee is that four of the six members —Playfair, Roscoe, Huxley and Lockyer—did more for the cause of the advancement of 'pure' and applied science than did any other four men during the period 1800–1914. Three of them had been teachers at the Royal School of Mines.

THE TEACHING PROFESSION

To summarise the position at the end of the nineteenth century: there was undoubtedly an increasing number of posts available for the scientific technologist; for the works chemist and analyst, the electrical engineer and electrotechnologist. But for the 'pure' chemist with research interests there were very few opportunities: the industrial research laboratories were then only in their infancy. For his colleague, the physicist, there were even fewer chances, for the first industrial physics laboratory had not then been founded in England.*

There is one other field that offers vocational opportunities for the professional scientist. This is the teaching profession; either in university or technical college, secondary or elementary school, and it was in this profession that the general employment of highly trained scientists first became common.

Broadly speaking, there were four series of major educational reforms in the closing decades of the century which were greatly to augment the vocational opportunities for the natural scientist. The first of these, in time, was the redistribution of school endowments under the Endowed Schools Act of 1867; this reform provided increased opportunities for the London and provincial graduate. A committee of London University had found, in 1879, that 'Among the graduates of the University of London, in the faculties of arts and science, there is probably a far larger proportion engaged in teaching than among those of the older universities. An increasing number of headmasterships in the schools reorganised under the Endowed Schools Act, and especially of science masterships are year by year obtained by London graduates'. Unfortunately, English secondary education was at that time very unsystematic, and statistics of the distribution of masterships do not exist. It is not, therefore, possible to do more than accept the university's statement at its face value.

Another field of reform was related to the growth of the technical college; the product of the great technical instruction movement which arose from Playfair's agitation. The effects of this on the educational

* Although Edison had founded one (in the U.S.A.) some twenty years previously.

landscape were well described by a Chief Inspector in an official report of the Board of Education issued at the very beginning of the twentieth century [30]. The inspector, C. A. Buckmaster, believed that: 'Nothing in English education is more remarkable than the manner in which special institutes for the purposes of science, art and technical work have sprung up all over the country during the last twenty years. When I began work as an inspector there were scarcely half a dozen buildings which had been erected primarily for the purpose of science teaching. Most of the classes attended by older students were held in mechanics' institutes, in mill or factory premises, in elementary schools or chapel buildings. In almost all cases the teachers depended upon the grants from the Department for their remuneration and it was largely owing to the enlightened self-interest of the teachers that the classes owed their existence'. This was a remarkable and apparently unsolicited tribute to the efforts and achievements of the pioneers of the technical education movement. Backed, as they were, by State, municipal and livery companies grants, these new colleges were able to offer teaching posts to the science graduates; as Playfair and James Hole had foreseen some thirty years before the movement commenced.

First in importance were the London polytechnics; in 1898, Sidney Webb [31] found that, at the Battersea Polytechnic, science graduates were doing original research and the laboratory enjoyed a grant from the Royal Society; while at Chelsea Polytechnic a Mr. Tomlinson, F.R.S., was running a research class. By that time some fifty thousand students attended the polytechnics and a number of these aspired to a London University degree. In the session 1896–7, one hundred at least were embarked upon post-matriculation science degree courses, and, in 1897, at least twelve London degrees were gained by polytechnic students.

The Royal Commission on Secondary Education* (1895) found that an increasing proportion of men and women teachers in secondary schools were drawn from the universities [32]. Thus of 720 students who left Newnham College between 1871 and 1893, 374 became teachers; while of 29 women graduates of Victoria University, 21 became teachers. In the second grade schools in the London area, a large number of the teachers were London graduates, and, in the third grade schools and organised science schools generally, most of the graduate teachers had been trained at either London or provincial colleges. The Commissioners found that of 2,958 public secondary

* The Chairman was James Bryce. Roscoe was among the members.

school masters, 1,459 had been trained at the older universities, 288 were London graduates, while the remainder were from provincial or foreign colleges. Of the 388 undergraduate masters, 267 were candidates of London University. It was also noticed that headmasters and those in authority increasingly demanded graduate teachers, and that competition for educational employment was now severe among graduates.

Of the 4,222 students at Oxford and Cambridge, 2,435 (58 per cent) came from the public schools. There were 1,440 scholarship holders but only 449 were from other than public schools. At Victoria, on the other hand, of 901 students, only 138 (15 per cent) were from public schools. There were 236 scholarship holders, but 204 of these came from other than public schools. Evidently, very few working-class students reached the older universities and not very many more got to Victoria. A very substantial hindrance was the paucity of secondary scholarships before 1900.

Two years after this Commission, a determined attempt was made to assess the state of secondary education in the country [33]. All types of secondary schools—endowed, subscribers, companies, local authority and so-called private-venture schools—were circulated and asked to give details of staff, pupils, etc. Combining the numbers of men and women teachers in boys', girls' and mixed schools, we find the following distribution of graduate teachers:

	Graduates	Visiting Graduates	Non-Graduates
Private Venture	2,884		
Subscribers	352		
Companies	498		
Endowed	2,320		
Local Authority	152		
Totals	6,206	1,938	17,136

making a grand total of 8,144 graduate teachers in secondary schools. This figure may be a little too large as it is possible that some of the visitors were also counted as regular teachers in other schools or might have been visitors to more than one school. It is, of course, extremely difficult to estimate how many of these were science teachers. In some cases the proportion must have been small; notably at the traditional public school. In other cases, as at the organised science schools, it must have been much higher: too high, perhaps, for a balanced curriculum. If we assume the arbitrary fraction of one-eighth —it can hardly have been less—then there would have been some 1,000 science graduates teaching in the secondary schools; and this is a number greater than those highly trained graduates who could have been employed in industry at that time. It is certainly much greater than the number who could have been employed in true applied science; in the industrial research laboratory.

It remains to add that, as a result of these findings and as a consequence of the initiative of men like Bryce, Sadler, Magnus, Roscoe and many others the reform of the secondary education of this country was taken in hand and, after the Education Act of 1899 and the Balfour Act of 1902 the new system of State secondary schools was set up.

Secondary and technical schools and colleges did not provide the only educational openings for the science graduate. He shares with his colleagues in the arts faculties the opportunity of university teaching and at this time the universities and university colleges were on the eve of their great expansion; an expansion enabled on the one hand by increasing State aid for higher education and, on the other, by the advancement of social welfare legislation.

The London colleges had, from the time of their foundations, drawn their professors of mathematics and natural philosophy chiefly from the ranks of Cambridge Wranglers. By 1877, Henry Latham was able to claim that the Tripos had become a professional education in mathematical physics and that high Wranglers were now assured of good scientific posts. Latham was referring to teaching posts in the new university colleges which, from 1870 onwards, were being founded in increasing numbers. Indeed, Latham had himself proposed, to the Devonshire Commission, the creation of new colleges, with consequent teaching posts, to increase the number of men willing to take university science courses. In later years, as more students were able to enter the universities, so the number of teaching staff was increased. Of course, the number was very small in comparison with those engaged in junior teaching, but the university teacher was able to carry out research and this was of great importance in the later development of applied science. It meant larger and better equiped laboratories together with research students working for the M.Sc. and D.Sc. degrees; and from such material potential applied scientists are made.

At the other end of the educational ladder the elementary schools now offered suitable teaching posts for science graduates. Two factors made this possible; there was the provision made in the code of 1890 for extended science teaching in the elementary schools and there was, in the same year, the recognition by the education authorities of the day training colleges attached to universities and university colleges. This was to be of great consequence in the peaceful years of the early twentieth century, affecting profoundly the development both of elementary schools and of universities.

Following the discovery by the Department in the seventies and eighties that the great national want was competent teachers of science, and the reorganisation of the South Kensington college with a view to overcoming that want, there occurred a sudden and sharp rise in the number of applicants for admission to that excellent college. The numbers of government (i.e. State scholarship) students rose from 25 in 1872 to 94 in 1885, when there were 136 fee-paying students, and rose again to 189 in 1895, when there were 119 fee-payers [34]. The proportion of scholarship holders was, therefore, far higher at the Royal College of Science than at any other comparable educational establishment. Not all these government students were teachers in training; there were also the Royal Exhibitioners and National Scholarship holders; but many of these, although under no obligation to do so, subsequently took up teaching. In fact in 1898 Norman Lockyer [35] estimated that not less than threequarters of all government-aided students became teachers. This would mean that, in 1895, there must have been at least 140 prospective teachers out of a total student body of 308; and as some of the fee payers must also have become teachers of one sort or another, it is at least plausible to conclude that the majority of R.C.S. students subsequently adopted that profession. In addition to these regular students there were the summer vacation courses for teachers. These were, at the end of the century, usually attended by between one hundred-and-fifty and two-hundred science teachers. Taking these figures together it is clear that the main function of the Royal College of Science was that of a Normal School.

Summarising the teaching profession, therefore, we conclude that the openings for the scientist were: (a) university teaching, the opportunities being few but yearly increasing in number; (b) secondary and (c) technical education; the first being complicated by the peculiar class structure of the public schools, but certainly the vacancies here were not inconsiderable, and, again, the number was increasing. As for technical schools born of the technical education movement, there too was an expanding field; (d) elementary education; either under the Department in the science classes, or under the educational authorities created by Forster's Act and hence in the Board schools.

The conclusion that can be drawn from these facts is that education provided the best opportunity for the highly trained science graduate. Applied research, either in industry or under government could not offer anything like the same number of posts, and it is very unlikely that industrial technology could offer any comparable opportunities to the 'pure' science graduate.

REFERENCES

[1] From the first wage-book of the Beyer-Peacock company; in the Manchester Museum of Science and Technology.

[2] *The Times*, 21 March, 1887.

[3] T. H. Huxley, Address at Manchester, 29 November, 1887. Published by the National Association for the Promotion of Technical Education.

[4] G. G. Ramsay, *Blackwood's Magazine* (1888) cxliii, pp. 425–43.

[5] Lord Armstrong (W. G. Armstrong), 'The Vague Cry for Technical Education', *The Nineteenth Century* (July, 1888) xxiv.

[6] Lyon Playfair, 'Lord Armstrong and Technical Education', ibid, (September, 1888) xxiv.

[7] Lord Armstrong, 'The Continuing Cry for Useless Knowledge', ibid, (November, 1888) xxiv.

[8] *Report of the Royal Commissioners appointed to consider whether any or what kind of new university or powers is, or are, required for the advancement of higher learning in London* (London 1889).

[9] *Report of the Commissioners appointed to consider the draft charter for the proposed Gresham University in London* (London 1894).

[10] Lord Playfair (Lyon Playfair), 'A Great University for London', *The Nineteenth Century* (October, 1895) xxxviii.

[11] M. E. Sadler and J. W. Edwards, 'Public Elementary Education', *Special Reports on Educational Subjects*, Cd. 8447 (1897).

[12] Auberon Herbert and others, 'The Sacrifice of Education to Examination', *The Nineteenth Century* (November, 1888) xxiv and ibid (February, 1889) xxv.

[13] H. E. Armstrong, *Report of International Congress on Technical Education*, Royal Society of Arts (London 1897).

[14] T. Mary Lockyer and Winifred L. Lockyer, *Life and Work of Sir Norman Lockyer* (London 1928) p. 189.

[15] P. Magnus, G. Redgrave, Swire Smith and W. Woodall, *Report on*

a visit to Germany with a view to ascertaining the recent progress of technical education in that country, C 8301 (1896).

[*16*] F. Rose, *Report on Chemical Instruction in Germany and the Growth and Present Condition of the German Chemical Industries,* miscellaneous series, Diplomatic and Consular Reports, Cd. 430–16 (1901). Supplementary Report, Cd. 787–9 (1902).

An interesting complement to Rose's reports is provided by *Reports of visits to foreign schools and institutions* (London and Manchester 1891). The visits were carried out by representatives of the Manchester Technical Instruction Committee and of the Manchester Technical School (now U.M.I.S.T.). Their report was illustrated with excellent photographs and plans of European technical colleges and thus provides visual confirmation of the relative backwardness of England at that time. I am indebted to Mr. Philip Short, of U.M.I.S.T., for calling my attention to this important work.

[*17*] John J. Beer, 'Coal Tar Dye Manufactures and the Origins of the Modern Industrial Research Laboratory', *Isis* (1958) il, pp. 133–34.

[*18*] Ibid.

[*19*] Ibid.

[*20*] Ibid.

[*21*] W. H. Perkin, Presidential address to the Chemical Society, *Journal of the Chemical Society* (1884) xlv, p. 219.

[*22*] *Vide supra,* pp. 104, 105. See also R. D. Welham, *J.S.D.C.* (1963) lxxix, p. 98 etc.

[*23*] D. S. L. Cardwell, 'The Development of Scientific Research in Modern Universities: a Comparative Study of Motives and Opportunities', in *Scientific Change,* edited by A. C. Crombie (Heinemann, London 1963) pp. 661–77. See also *Raphael Meldola,* edited by J. Marchant (London 1916) pp. 4, 65.

[*24*] Conrad Gill and Asa Briggs, *The History of Birmingham* (2 vols., Oxford University Press 1952) i, p. 109.

[*25*] From a confidential report in the archives of the University of Leeds.

[*26*] C. M. Mellor, *Dyeing and Dyestuffs, 1750–1914.*

[*27*] Lord Rayleigh, *J. W. Strutt: third Baron Rayleigh, O.M., F.R.S.* (Edward Arnold, London 1924) pp. 276–85.

[*28*] *Report of the Committee Appointed by the Treasury to Consider the Desirability of Establishing a National Physical Laboratory,* C 8976–7 (1898).

[*29*] For an account of the Bishop Berkeley Fellowships, see E. Fiddes, *Chapters in the History of Owens' College and of Manchester University* (Manchester University Press, 1937) Appendix.

[30] Report of the Board of Education, 1900–1901, ii, p. 248.

[31] Sidney Webb, 'The London Polytechnic Institutes', Special Reports on Educational Subjects, Cd. 8943 (1898) ii.

[32] Reports of the Royal Commission on Secondary Education, C 7862– (9 vols., London 1895) i, p. 238; iv, Appendix G and ix.

[33] Return of Secondary Schools in England, Cd. 8634 (1897).

[34] 20th, 33rd and 43rd Reports of the Department of Science and Art.

[35] J. N. Lockyer, 'A Short History of Scientific Education', a lecture given at the Royal College of Science, 6 October, 1898. Printed in Education and National Progress.

CHAPTER SEVEN

The New Century:

1900—18

The great International Exhibition of 1851 was a triumph for British industry and technology. Had, however, the Royal Society of Arts decided to hold a similar exhibition in London in 1901, the overall impression would have been far less satisfactory as far as Britain was concerned. For example, the internal combustion engine—the new 'heat engine' that was to become one of the most important motive agents of the twentieth century—was to a great extent a Continental development in all its forms. Just as the names of Newcomen, Watt, Hornblower, Trevithick, Murray, Woolf, Stephenson and McNaught had indicated the main stages in the development of the steam engine, so those of Beau de Rochas, Lenoir, Otto, Langen, Maybach, Benz, Daimler and Diesel marked the advances of the internal combustion engine. Further, the first really important applications of the new heat engine—to motor cars and, very soon, to aircraft—were due to the genius of French and American engineers. In the other and related key technology, that of machine tools, England had already been overtaken by Germany and America.

The success of the German synthetic dyestuffs industry, which we have already discussed, led by a simple 'chain effect' [1] to radical developments in other fields of science, applied science and technology. Thus Paul Ehrlich began his career by studying the use of the new dyestuffs for staining tissues and then went on to examine their biological and particularly therapeutic effects: a line of inquiry that ultimately led to such discoveries as that of salvarsan. Ehrlich's work was a contributory factor in the rise of the great German pharmaceutical industry at this time. But besides this, the stimulating effect of the synthetic dyestuffs industry extended back, as it were, to the heavy chemicals industries which supplied the basic raw materials for the new, science-based enterprises. In other fields, too, in glass technology, scientific instruments, cameras and a wide range of electrical components, Germany had outstripped Britain. In fact we can sum up the situation briefly, using Francis Bacon's classification: in the sphere of science-based innovation the honours lay with the methodical, scientific German; in empirical technology and production engineering, with the enterprising and resourceful American.

H. J. Habbakuk has pointed out [2] that, in contrast to Britain, a shortage of skilled labour coupled with high wages and an abundance of cheap land awaiting settlement meant that in America there was every incentive for the invention and development of labour-saving and mass-production machinery. L. T. C. Rolt has underlined the technological gap between the two countries by his comment that while the early Ford cars, much looked down on by old-fashioned Englishmen of the day, represented a new concept in mass-production technology; the Rolls-Royce, much admired by the same Englishmen, manifested a very high standard of conservative craftsmanship for the benefit of the wealthy few.

To these reasons for the supremacy of American empirical technology we can perhaps add such factors as a more enterprising population with a high standard of general and practical education, fewer social barriers and a sensible political constitution. As far as Germany was concerned, besides superior education at all levels, the contributory causes that have been suggested include relatively liberal patent laws and an absence of heavy capital investment in old-established industries—one of the advantages of *not* being first in the field—together with certain specific factors, such as freedom from burdens like the alcohol duties that were levied in Britain. But it is improbable that these economic factors were as important, individually or collectively, as education and technical skills.

A significant pointer in this direction was the number and quality of the foreign experts who came to England and established successful firms based on the new technologies. Young Germans and Swiss had, as we saw, been willing to seek employment as scientists and technologists in England. But towards the end of the nineteenth century the pattern seems to have changed: the immigrants were no longer employees; they had become employers, entrepreneurs and pioneers of the new science-based industries. In this important group we find such people as Ludwig Mond, Charles Dreyfus, Hans Renold, William Siemens, Heinrich Simon and Guglielmo Marconi. Others, like the Mannesmann brother and Henry Ford, established factories in England without themselves settling in the country [4]. Inevitably one is reminded (uncomfortably) of the state of affairs in the early seventeenth century when those responsible for technical innovation in England were immigrant, or visiting Frenchmen, Germans, Italians and Dutchmen.

The withering of British enterprise was quite general. We have already seen that chemistry in Britain had been in a debilitated state during the second half of the nineteenth century. But physics and

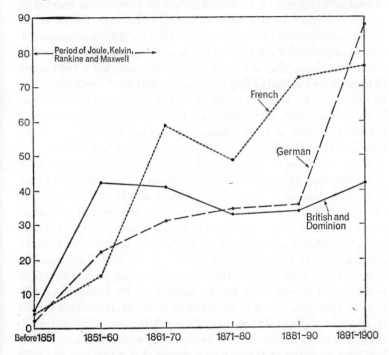

Fig. 1 Papers relating to heat and thermodynamics appearing in British, German and French periodicals, 1851–1900. Source: *The Royal Society Index*

mathematics were no healthier as the century came to an end. We can demonstrate this very conveniently by considering British contributions to thermodynamics [*5*]. This is a branch of science that began with the study of the performance of steam engines invented and developed by British engineers and to the formal establishment of which British scientists had made vital contributions.* Figure 1 shows the number of papers on thermodynamics published per decade in British, French and German periodicals between 1850 and 1900.

* i.e., Watt (considered as a scientist), Robison, Southern, Davies Gilbert, Dalton, Herapath, Forbes, Joule, Rankine, Kelvin and Maxwell.

The data is derived from the Royal Society Index of Scientific Papers [6]: a source unlikely to overlook British contributions. Nevertheless it will be seen that after a promising beginning the number of British publications remained more or less static for about fifty years while those of Germany and France rose; the former very rapidly, the latter at least satisfactorily. A continued failure to increase the number of papers published in this field indicates, when such things as the population increases are considered, an effective decline of science. It may be objected that a mere count of heads is no safe guide to creative scientific activity. This is true in specific instances, but a comparison of trends over a period of fifty years can hardly be without significance. In any case if we consider quality rather than quantity British failure is no less apparent. The notable contributors to thermodynamics in the last two decades of the nineteenth century and the first of the twentieth were Continentals and Americans;* none was British.

What was true of thermodynamics appears to have been true of most other branches of physics and mathematics; although fundamental work was done in electromagnetic theory by three Irishmen, Fitz-gerald, Larmor and Stoney. Accompanying the general decline in science and technology there appears to have been a disinterest in such indicative subjects as the history of science, the history of technology and the study of the social implications of technological change. The economic historian Arnold Toynbee, for example, hardly mentions technological factors in his famous lectures on the industrial revolution [7]. One man, however, did write an excellent study of the history of scientific thought during the nineteenth century. This was J. T. Merz, an industrial chemist who lived on the north-east coast but who had been educated at German universities. His work was not fully appreciated at the time and only now is its true value being realised [8].

The conclusion one must draw from this is simple. Economic factors might well account for a decline in industrial technology but they can hardly account for the failure of abstract science that took place at the same time in what was still the wealthiest country in the world. And since science, abstract science and technology are undeniably linked to one another it must follow that economic factors cannot wholly, or

---

* Cf. Boltzmann, Wien, Stefan, Planck, Nernst, Van't Hoff, Van der Waals, Berthelot, Le Chatelier, Duhem and, of course, Willard Gibbs. The names of the English contemporaries should be sought—and their contributions assessed—from the pages of standard text-books on heat and thermo-dynamics.

perhaps even in the main account for the decline in the latter. Education is the common denominator, but before we proceed further we must examine two hypotheses that may help to explain the *fin de siècle* lassitude of British science.

The first hypothesis relates to the ideology, characteristic of the period, which stands in sharp contrast to the prevalent ideals of the heroic age of industrialism that was ending in 1851. This was the raucous creed of Imperialism that may be said to have been initiated by Disraeli's declaration of the Queen as Empress of India (1876). From many points of view the ideals of the age of Imperialism seem much more remote from the present day than do those of the earlier age of industrialism. A rush to paint the map red, to establish dominion over countless square miles of jungle, desert and swamp was surely to mistake the outwards appearances of power for its essential inner structure. How effective all this was in diverting the energy and ambitions of able young men from the pursuits of technology, industry and perhaps even of science can only be conjectured, for the roots of Imperial ambition must have reached far back into the public schools where so many of the nation's middle-class youth were being educated. How many young men from such schools would seriously have considered a career with, for example, Beyer-Peacock or Brunner-Mond?

The second, and related hypothesis, raises a disturbing question: were the old radicals right after all? Did the intervention of the State in education have a deleterious effect on the pursuit of science? The poverty in numbers and the limited resources of British universities in 1901 have been sufficiently contrasted with the advantages enjoyed by German and American universities. But the simple arithmetic of student numbers and State subsidies may well conceal another and more fundamental defect. At the end of the nineteenth century the British universities, like the German universities before them, were making what turned out to be a highly successful take-over bid for science, and scholarship generally. So successful have they been in this that it is difficult for us now even to imagine a time when scientists were not necessarily men with university degrees and when a high proportion of the country's scientific research was not done within university laboratories. Universities have many virtues, but they have faults, too, and it is at least possible that they imposed some of their faults on the science they appropriated: excessive specialisation being a case in point. Another possible fault can best be explained by invoking one or two simple biographical examples.

The year 1970 marked the bicentenary of Captain James Cook's landing in Australia. The great navigator's father and mother were farm servants and he began his career as an ordinary seaman on a collier. Nevertheless he rose to be a Post-Captain in the Royal Navy and a Fellow of the Royal Society before his death at the early age of 51. The intriguing point is that, notwithstanding nineteenth century reforms, a man with Cook's social background could not have been commissioned in the executive branch of the Royal Navy of 1901. Indeed, Professor Marder has shown that by this time commissioned rank had become the preserve of the wealthy to such an extent that Nelson himself would have been excluded from the Navy [9]!

In fact, executive officers of the Royal Navy were drawn, Professor Marder indicates, not from a national population of some forty million but only from that section of the population, amounting to about a million, that could afford the heavy expenses of education before and during naval college. The discrimination based on money that was thus imposed was not a matter of deliberate policy. It had developed over the years and no one, in effect, had noticed what was happening. When we recall the scientific aspect of Cook's work, we cannot help wondering whether the same sort of discrimination applied later on to careers in science as it did to careers in the Navy. It seems, *a priori*, quite probable that it did. After all, there is no reason to suppose that dons and university authorities in general are or were more liberal in spirit or democratic in intention than admirals or admiralty officials.

If, in order to win recognition as a scientist a man was required to have gained a university degree it becomes a matter of some interest to know from what proportion of the population the graduates of 1901 were drawn. Secondary education was not free, still less were university courses. Indeed, it seems very probable that the population from which the scientists of 1901 were drawn must have been much less than the total national population of forty million. The number of people who could afford to keep a son—or daughter—at secondary school and then at university to be followed perhaps by the expense of post-graduate research training may well have been five or at most ten millions. It is salutary to recall, at this point, that two of the ablest and most influential of British scientists of the nineteenth century—John Dalton and Michael Faraday—came, like Cook, of very humble parentage.* Had

* Among other scientists of this and the previous generation who were also of working-class origin were Sir Humphry Davy, Sir John Leslie and Sir James Ivory, the mathematician.

they been born during the last three decades of the nineteenth century could they have managed to secure a university education and thereafter could they have afforded to embark on research work? It is, at least, problematical. And if there is some doubt in the cases of Dalton and Faraday how much more must there be in the cases of the able second rankers on whom the advance of science is also dependent? I would therefore suggest that a rewarding investigation for an historical sociologist of science might be a Namierite study of the social origins, education and subsequent careers of the Fellows of the Royal Society who were elected over the years 1880 to 1900.

Late eighteenth- and early nineteenth-century England had many faults: it was a cruel, hypocritical and often oppressive country in which great extremes of wealth and poverty existed side by side. But in one respect at least it could claim to be an open society: a young man with genuine scientific ability could, if he was fortunate enough to survive the physical hazards of infancy and youth, rise to the top of the scientific tree. Early twentieth-century England, however superior it might have been in other respects, may well have lacked this ultimate, saving grace.

Observations such as those above may seem no more than the wisdom of hindsight, especially as the Edwardian age is often regarded as the golden summer of British power and civilisation before the disastrous era of great wars began in 1914. But even contemporary observers suspected that, by 1900, Britain had slipped badly in the struggle for existence. The humiliations of the Boer War, on the national scale; the criticisms of H. G. Wells and other interpreters of the social scene together with the gloomy prophecies of Karl Pearson and the eugenists on the subject of national degeneracy, do not leave that happy impression on the mind that present nostalgia tends to evoke.

On the credit side, however, there was a loosening of dogma, a growing recognition on the part of the reformers that the problems of science and technology could be tackled in a scientific manner and were not merely a question of general and futile exhortations. Lockyer, writing in 1901, compared our position at the beginning of the new century with what it had been in 1801, at the outset of the railway age— now, the chief London electric railway was American. He asked the question, as a good scientist should: ' . . . is there no scientific method open to us to get at the real origin of the causes which have produced the present anxiety?' Pointing to the German educational system he remarked that: 'Here we tried to start chemical industries practically without chemists, as Mr. Perkin has told us. In Germany they are now carried on by scores, in one case more than a hundred, of the best trained chemists the country can produce, in research laboratories attached to all the great works' [10].

The paradox that England, the first country to be industrialised was the last to achieve national education, was stressed by the contemporary observer Fabian Ware [11]. The reason Ware adduced for this: the struggle between oligarchy and liberalism, the absence of national challenge and so of national purpose, and the current doctrine of self-interest may well be valid; but they do not tell us in what way the educational endeavour failed. To this question he returned exactly the same answer as Matthew Arnold: that it failed in that it did not diffuse a knowledge of science among those who controlled the new forces introduced in industry by the discoveries of science; and this was due to the inability to cope with the problem of secondary education, conceived not as technical instruction, nor as formal classicism, but as an education fitted to the natural development of the pupil. The answer for Ware, as for many others, lay in the secondary school. It did not lie elsewhere, for we had primary education and we had technical education—more, in fact, than the Germans had.

Arthur Shadwell was another writer who wanted to soft-pedal the technical education movement and to direct attention to the production of 'the officers, not of the rank and file', as they had done in Germany. For the future we must look to the newer universities and the recent science departments at the older foundations, notably Cambridge. 'With the universities, the National Physical Laboratory and the coming Imperial College at South Kensington, it is not the schools that we lack now, but the scholars' [12]. Shadwell was a great admirer of the 'scientific' Germans; he ascribed their industrial success to the foresight of their manufacturers who had long ago realised the value of highly trained men in industry and had so created a demand for them. He did not try to find out, still less to explain, why the Germans may have been so enlightened; and it can hardly be said that his observations were in advance of those current among scientists and industrialists forty or fifty years before.

It was generally recognised that the new century would call for new techniques and new ways of thought. While *The Times* and other papers were discussing what they felt to be an alarming decline in national intellect, Joseph Chamberlain was asking for new Kelvins [13] and for more universities as vital ramparts of our industrial defences [*14*]. The physiologist E. H. Starling also demanded the foundation of new universities when he pointed out that future competition would be 'not for lands but for control over the forces of nature'; for the sake of industry 'pure' science must be advanced [*15*]. In the same context Arthur Balfour had asked the questions: 'We, the richest country in the world, lag behind Germany, France, Switzerland and Italy. Is it not disgraceful? Are we too poor or are we too stupid?' [*16*]. But the issue was put most succinctly by Sir William Huggins, in his 1902 Presidential Address to the Royal Society [*17*]. Granting that there was little demand for applied scientists and technologists and that the leaders of the nation were mainly ignorant of science, Huggins inferred that this must be due either to a defect of character or to a defect of education. He concluded quite simply that it must be due to the latter.

The next challenge was made by Norman Lockyer in his Presidential Address to the 1903 meeting of the British Association. Outlining German success, he deduced as the causes of our failure, the insufficiency of universities, the neglect of science in schools and universities and the consequent ignorance of it in the Civil Service and in commerce. The remedy was the State endowment of universities and the foundation of at least eight new ones. At the same time he propounded his educational theory: it is not instruction we want, it is research: for research is the most powerful instrument of education we possess. He had, he continued, been thinking for some time about the desirability of founding a chamber or guild to advance the cause of science; but he had recently discovered that the British Association itself was, in the first instance, founded for that very purpose (see p. 60). Therefore he believed that the British Association should take upon itself the burden of science as a national interest.*

Another active propagandist at this time was Philip Magnus. He had an interesting contribution to make at the Oxford summer school of 1903 [*18*]. 'For the successful application of science to industry it is essential that a country should possess an army of highly trained and intelligent scientific men capable of directing the different kinds of engineering work, and of constituting what is so much needed—an intelligence department to every factory.' Magnus certainly had the modern idea: 'before long the laboratory will be regarded as a no less

* Although he was not able to persuade the Association to act as a public sponsor of science, he did, as the next best thing, succeed in founding the British Science Guild.

important part of a factory than the drawing office or counting house'.

Also arising from these discussions was the action of the L.C.C. Technical Education Board which, in 1902, initiated an inquiry* into the problems affecting the application of science to industry and to throw light on the way in which certain industries, especially in the London area, had been lost to foreign countries. A large number of scientific witnesses† were examined and a report issued in the July of that year.

The conclusion was that England's relative failure was due to the defective educational system and to the superiority of foreign scientific education. This was most marked, in its effects, in the chemical, optical and electrical industries, and there were, it was thought, four particular causes to account for this:

(1) Lack of scientific training of manufacturers resulted in inability to understand the value of science.

(2) Bad secondary education meant that few were really fit to receive advanced technological education.

(3) An insufficient supply of young men properly trained in science and in the techniques of applied science.

(4) An absence of a higher technical institution sufficiently well endowed to enable it to give adequate attention to postgraduate and advanced work.

But the worst deficiency, according to the report, was in secondary education: 'In the majority of secondary schools the curriculum has been so hampered by the exigencies of examining authorities and of examinations, that the teacher has been compelled to devote undue attention to storing the minds of the students with facts for reproduction at the expense of the time which should be devoted to stimulating their reflective powers and making them think. In after life those who enter upon industrial pursuits too often regard science with distrust,

* The inquiry was conducted by officers of the L.C.C. aided by Mrs. Bryant, Sir Philip Magnus, Sir Owen Roberts, Graham Wallas and Sidney Webb.

† J. F. Swann, B. Samuelson, Roscoe, F. Clowes, J. Dewar, J. T. Merz, W. H. Perkin, W. Ramsay, T. Tyrer, R. Meldola, G. T. Beilby, T. H. Thorpe, W. E. Ayrton, H. E. Armstrong, I. Levinstein, A. Siemens, H. Jackson, H. Bell, A. Rücker, A. H. Green and G. Parker, with written reports from M. J. M. Hill, Prof. Cormack, Prof. A. Fleming, Prof. Lünge and B. Seebohm Rowntree.

and to some extent this distrust is merited, owing to the insufficient preparation and training of those who offer themselves for responsible posts in scientific industries'.

The London university colleges were hampered by lack of endowment, accommodation, teaching power and equipment and by the inadequate preparatory training of the students on entry. It was believed that the highest grade of technical instruction should be in a day institution and not in the evening polytechnic.

The main need in London, the report continued, was for co-ordination of the highest grade teaching and post-graduate work. For those who were going to be leaders in scientific industry a course of education was strongly recommended which would include a general secondary education, classical or modern, up to seventeen or eighteen, to be followed by three years' work for the B.Sc. degree. In the meanwhile England continued to be held back by inferior secondary education and the lack of facilities for advanced training and research.

Together with the industrial and university chemical laboratories, one of the most admired of German institutions at that time was the great polytechnic at Charlottenburg. In July 1903 the flattery became quite apparent when Lord Rosebery wrote to Lord Monkswell [19] Chairman of the L.C.C., outlining a scheme for a London 'Charlottenburg' and promising aid from private individuals if the L.C.C. would give its support. The Council approved, and a sub-committee was appointed to draw up a scheme to include the Royal College of Science and the Royal School of Mines, and the City and Guilds Institute Central Technical College. The Council offered £20,000 a year grant and, to explore the possibilities, a Departmental committee under Sir Francis Mowatt, and later R. B. Haldane, sat for two years (1904–6). In their report they recommended the fusion of the South Kensington Colleges and the building of new premises. The Board of Education agreed to hand over the two State colleges which thus underwent their final mutation to become the Imperial College. This may be described, not unfairly, as the third attempt to found a technical university.

In 1908, a Board of Governors was elected, a Rector appointed and the Charter of the Imperial College granted. In March of the following year a Royal Commission was appointed to consider the relationship that was to exist between the new college and the recently reorganised University of London, and also to consider the facilities for research in London. For four years this Commission examined London University with a zeal and thoroughness that that university had been wont to extend to the candidates for its honours. They did not limit themselves to Imperial College, but investigated every mode of education in London above the school level, including the medical colleges, legal education and the polytechnics. While the results were hardly in proportion to the effort, it was decided that Imperial College 'founded to give the highest specialised instruction and to provide the fullest equipment for the most advanced training and research', must be the London Charlottenburg [20]. The absence of definite technical instruction in England caused 'a want of co-ordination between science and industry. That want must be corrected by Imperial College'.*

UNIVERSITIES—THE CONTINUED TREND TO SPECIALISATION

It is sometimes suggested that the present 'narrow' specialism of degree studies is the consequence of the intrusion of science in the university syllabuses, the demands of industry and the rise of applied science, which developments, it is implied, stand in sharp contrast to the older educational traditions of broad and liberal studies. Enough has been said above, I think, to show that such views are oversimplifications of a complex matter. The trend to specialism was clearly apparent long before these changes and 'reforms' has taken place, before 'natural'—or progressive—science had ever entered the educational syllabus. While it was perfectly possible, indeed customary, to enjoy a liberal education in the universities and colleges of a century ago, the greatest honours went to the specialist. Indeed, the very name and certainly the nature of the long established 'Honours' system imply specialisation.

When we last considered the London B.Sc. the requirement was that, at the second B.Sc. stage, the candidate should select any three of the natural science subjects or logic and moral philosophy after which, having passed this examination, he could take honours in his chosen subject—mathematics, chemistry, physics, etc. By 1900 he could, at Intermediate level (as the first B.Sc. examination was now called) select three subjects and, if a physical scientist, was under no obligation to pass in general biology. The Pass degree still required three subjects as before, and if Honours were taken it was necessary to pass in two other subjects. By 1910 the course had again been altered: the interval between Intermediate and Final had lengthened and, while the Pass degree still required three subjects, the Honours candidate was

* Cf. Lyon Playfair (p. 88).

obliged to follow a different course, studying his principal subject together with an approved subsidiary. Thus the originally liberal London science degree had gradually evolved until finally it became a highly specialised qualification. It may be accepted that the other universities followed more or less the same pattern. The peculiar irony is that the much feared Dutch-auction of degrees did not take place; in fact, the opposite occurred; there has, if anything, been a competition upwards.

Naturally questions continued to be asked as to whether this was the right policy. Sir Norman Lockyer, like H. E. Armstrong, continued to be a bitter and unrelenting critic of examinations; for Lockyer, of course, research was the mode of education. He was not alone in maintaining this, for Sir William Huggins, O.M.,* Lockyer's fellow astronomer, had demanded that teacher and student be less 'shackled' by the hampering fetters of examinational restriction. Students, according to Huggins, should have greater freedom to learn and to do research and in this respect he commended the post-graduate Johns Hopkins University. The dry bones of academic reading and examination need the living breath of research, the mind must find its own way best suited to its powers: 'The creative use of imagination is not only the fountain of all inspiration in poetry and art, but is also the source of discovery in science and, indeed, supplies the initial impulse to all development and progress'.†

Three years later Sir Arthur Schuster could say that 'It is much easier to teach if you make the accumulation of knowledge the primary object, and it is so difficult to test by examination anything except the possession of knowledge. . . .' Disliking the competitive element induced by written examinations and wanting to use research as a means of higher education, Schuster propounded his philosophy of university education in the following terms: 'The true function of a university can only come into play when the student is made to work in a restful spirit which excludes anxiety . . .' [21]. A rather different point of view from Schuster's was taken by Sir Oliver Lodge, who greatly disapproved of the disintegration caused by highly specialised studies. Lodge wanted a general cultural education rather than the high specialisation by then achieved; an ideal with which Sir William Ramsay agreed [22].

* Huggins was, with Kelvin and Rayleigh, one of the first to be awarded the O.M.
† Presidential Address to the Royal Society, 1902.

Some reconciliation of these positions was attempted by Philip Hartog [23]. Lodge's general course, Hartog maintained, could best be achieved at the secondary school. But secondary education is defective because the secondary teacher has very rarely had any research experience and because he is overworked, underpaid and insecure. Examinations can be considerably improved; for example, papers which award equal marks for essential and unessential questions, and have pass marks of only 30–40 per cent are bad, but these flaws can be remedied to a great extent. For most science men of between 18 and 22 years of age, research work is desirable; if you do not give them time to do this you may sterilise their original faculties utterly. And, continued Hartog, 'the attitude of continually working to please other people instead of working to please and satisfy one's own mental requirements and critical power, is the attitude systematically encouraged at the present day by English secondary and university education, taken as a whole'.

Six years later Hartog returned to the question of examinations. In an interesting and carefully reasoned paper, read before the Society of Arts,* he likened the examination system to an artificial nervous system controlling our educational institutions [24]. The competitive element leads us to forget the distinction between the efficiency of an examination and its difficulty. Examination enthusiasts say, 'This examination is better than that for it is harder to pass'. Should we not, reasoned Hartog, rather ask 'Is the man who passes it a more useful member of society?' This question is most important when we come to deal with technical examinations: when we are concerned with the training of teachers, lawyers and industrial chemists. When we come to consider 'culture' it is very possible that sensitiveness and responsiveness suffer injury from the intellectual repression required for the examination syllabus: 'at every step the feelers of the mind are paralysed by the suggestion, "I am wasting my time in going farther; that won't be asked". It may be held, and I should agree, that culture is as individual a thing as conscience; that culture may be killed, that it cannot be caught, by examinations.' Hartog's investigations of examinations left him unsatisfied, and he felt justified in calling for a Royal Commission to investigate the problem. In this he had the support of Dr. H. A. Miers (Principal of London University), Professor John Adams, Sir Arthur Schuster and Michael Sadler. This was the second time a Royal Commission on Examinations had been called

* Lord Curzon took the Chair.

for; the first occasion was during the agitation of 1888; but in neither case was the demand fulfilled.

Thus on the eve of the Great War there were many men, and some of them very eminent, who had criticised or were criticising, the written examination system as applied to higher education. Among the chemists Roscoe, Ramsay and Armstrong; among the physicists Kelvin and Schuster; among the astronomers Lockyer and Huggins had all at some time or another expressed the gravest doubts as to the efficacy of the method of high-level written examination. Why then did the practice continue, and not only continue but steadily spread in extent? The answers to this are probably many, but tentatively we may note: in the first place it was associated with democratic tendencies and was in accordance with received ideas of social justice. One may accuse a selection board of favouritism with regard to class, sex or religion, but you cannot accuse an examiner who patently does not know whose paper he is marking. In the second place, as already remarked, it provides a considerable incentive to sustained study and, in the third place, it is economical in both time and effort: one paper can 'test' many candidates. With regard to university science education the substitution of research degrees as so frequently demanded would have necessitated a much greater endowment in order to meet the staff and laboratory requirements. That, of course, was not forthcoming in spite of the urgent demands of Lockyer and others [25]. There was also the administrative difficulty; the main university institutions were, by now, wedded to their time-honoured examination systems and there was also the London External degree to complicate the issue.

The practice of specialisation was not limited to written examinations; it had in fact, even wider ramifications. Although the Royal Society had largely been composed of natural scientists it had not generally excluded distinguished philosophers and social scientists. However, events were now to force a decision; in 1899 what was to be the first of a series of regular international conferences of academics was held at Wiesbaden and it was then discovered that most foreign academics had sections devoted to the human sciences. The questions therefore arose: what British organisation was competent to represent philosophy and the social sciences at future conferences? Should the Royal Society so enlarge its scope as to include those studies? A committee of the Society was set up and evidence taken from the representatives of the human sciences. Apparently these latter set a high value on inclusion within the Royal Society for there was general agreement among them that it was very desirable. But, after a special meeting of the Royal Society, on 9 May, 1901, inclusion was rejected and the 'excluded' thereupon decided to form the British Academy; incorporation being petitioned for on 17 December of that year. This much displeased the liberal-minded Lockyer, who complained that '. . . subjects the study of which by scientific methods increase the sum of natural knowledge must all stand on the same footing. I use the word "scientific" in its widest, which I believe to be the truest, sense, as including all additions to natural knowledge got by investigation. Human history and development are as important to mankind as is the history and development of fishes. The Royal Society now practically neglects the one and encourages the other' [26].

THE UNIVERSITIES—FINANCES AND EXPANSION

Norman Lockyer's Presidential address to the British Association meeting at Southport in 1903 suggested, by its theme and title,* that he did not intend to be a mere figurehead during his year of office. It was largely through his efforts that, in the following July, a high-powered deputation waited on the Prime Minister, Arthur Balfour, with a reasoned request for bigger endowments for universities and colleges. The deputation was, in the event, well-timed; it was also rather obviously timed for Balfour—a very unusual senior politician in that he also had an understanding of science†—was to succeed Lockyer as President of the British Association in about three months time. The case for increased endowments was accepted and it was promised that the grant should be doubled and, it was hoped, doubled again in the following year. Although this was far less than Lockyer had hoped for, it was at least something, and from that date onwards the grant for universities has steadily risen. Set out below are the figures for the English universities and, for comparison, the State subventions for the Prussian universities:

Year	English	Prussian
1897–98	£26,000	
1900–01		£476,000

* 'The Influence of Brain Power on History'.

† Balfour's Presidential address was concerned with the latest, sub-atomic theories of matter. It is worth recalling that his uncle, Lord Salisbury, who had also been Prime Minister, was interested in science.

Year	English	Prussian
1902–03	£40,000	
1904–05	£69,000	
1905–06	£115,000	
1906–07		£588,000
1910–11	£123,000	£700,000
1913–14	£170,000	

By the outbreak of war the grand total of public (including municipal) aid for the English universities was about £250,000; and although this was barely one-third of the annual grant for the Prussian equivalents the increase of endowment was certainly substantial. But the effect of this on student attendance etc., is difficult to estimate, for many of the colleges had very long straggling tails of part-time and evening students. Limiting ourselves, therefore, to full-time students, and even this term is rather elastic, there were in 1900 about 4,500 full-time students. Before the outbreak of war the number had risen to about 9,000, although for about three years before 1914 the trend of expansion had reversed and numbers fell a little. While the numerical increase was not in proportion to the increased expenditure, the rise in standard in the quality of work achieved certainly was: approximately four times as many students were enabled to take degrees. This process was expedited by the formation of the University of Birmingham (1900) and the break up of the federal Victoria University into the civic universities of Manchester, Liverpool and Leeds (1903); to be followed by the founding of Bristol and Sheffield universities. In effect this meant the final overthrow of the examination-board university and the unrestricted rule of the examiner.

Yet these figures, promising as they were, did not indicate that England was making progress against German 'competition'. The German student body had increased from 27,000 in 1893 to 33,000 in 1899, and to 58,000 in 1910. At the same time the polytechnic students had increased from 11,000 in 1899 to 16,000 in 1910; and, as the polytechnics now awarded degrees, these students could properly be counted with the universities' population. The numbers of students taking comparable courses in higher technology in England at the two latter dates have been estimated at 2,000 and 4,000 respectively. Unfortunately the English numbers include many who were at universities and have, therefore, already been included in the count of full-time students.

It was calculated that, in 1900, the number of day students per 10,000 of the population was:

In the U.K.	In the U.S.A.	In Germany
5.0	12.8	7.9

—a proportion not too discreditable to the United Kingdom. But it should be remembered that the excellent Scottish universities would naturally tend to raise the average, and that the education of the German student was almost certainly much more systematic and advanced than that of most of his British colleagues.

VOCATIONAL OPPORTUNITIES

The number of degrees awarded by the university colleges and, for that matter, by Oxford and Cambridge, rose very steeply in the years 1900–14 (see Figures 2 and 3). The point now to be discussed is: what use was made of their qualifications by those who took science degrees? A question that brings us to a consideration of the industrial development of applied science. It is interesting to note, *en passant*, that in 1902 an investigating committee of the Calico-Printers Association recommended the formation of a central research department. This was most strongly urged by Sir William Mather [27] who added a report of his own, setting out the advantages of such an establishment. His notions were thoroughly progressive in that he wanted a research and experimental laboratory, well-equipped and conducted by the ablest and best-trained chemists. Unfortunately this scheme fell into abeyance.

The question of the number of scientific chemists employed in British industry was raised at the 1902 British Association meeting. A small committee comprising Professor W. H. Perkin (jun.), Professor G. G. Henderson, Professor H. E. Armstrong and G. T. Beilby was set up to try to find the answer. They circularised members of the Society of Chemical Industry of which the great majority of technical chemists in the country were members. More than half replied and those who did not were, they concluded, probably not engaged in chemical works. Therefore they felt that their statistics gave a fair idea of the situation [28].

Of the 502 chemists they counted, only 107 were graduates. These were further classified as follows: 59 were graduates of British uni-

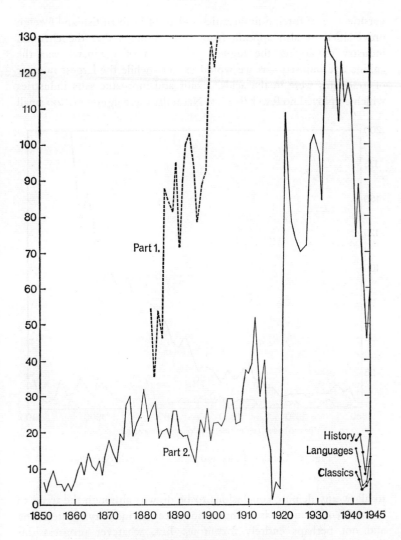

Fig. 2. Cambridge Natural Sciences Tripos, 1851–1945

versities; 32 of foreign universities and 16 of both British and foreign universities. It also appears from their findings that the chemical industry employing the highest proportion of graduates was the aniline dye industry—as we would expect—while the largest number of graduates were in the acids, alkalis and inorganic salts industries which employed no fewer than 17. Naturally these figures are too small

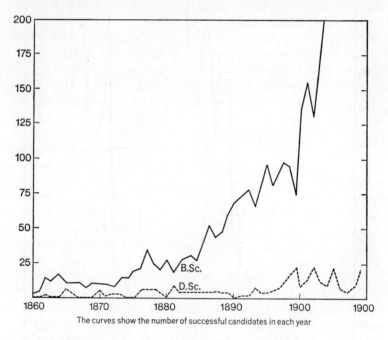

The curves show the number of successful candidates in each year

Fig. 3. London B.Sc. and D.Sc. Degrees from 1860

to allow any definite conclusions to be drawn; although the trend to more scientific methods in the aniline-dyestuffs industry is interesting and not perhaps entirely fortuitous. But, whatever progress this distribution of graduate chemists represented, the 107 graduates do not compare with the numbers in German industry.* Of these 107 graduates, no fewer than 48 had foreign degrees which, we may infer, were almost all German or Swiss. This shows that even at that late date, the professional scientist in industry was to a large extent a foreign product.

* Cf. Dr. Rose.

This matter was referred to by Sir James Dewar in his Presidential Address to the Association. Using the data of the committee he estimated that, at a very liberal allowance, there were 1,500 chemists in industry. The great majority of these would have had little training and would, as Sir William Ramsay later put it, be paid a labourer's wages. In any case this was but one-third of Germany's total of 4,000 chemists, and Dewar further calculated that, while 84 per cent of the German chemists had been trained either at university or polytechnic (74 per cent and 10 per cent respectively), only 34 per cent of the British chemists were so qualified. The German chemists,* Dewar believed, were better trained; their degrees were awarded for work done, not for 'questions asked and answered on paper'. In fact, 'There are plenty of chemists turned out, even by our universities, who would be no use to Bayer & Co. They are chock full of formulae, they can recite theories and they know textbooks by heart; but put them to solve a new problem, freshly arisen in the laboratory, and you will find that their learning is all dead.' We have, Dewar told his audience, to begin at the beginning and train people to solve problems for themselves 'instead of learning by rote the solution given by somebody else'. The main trouble was, 'I give it in a word—want of education'. England had material, capital, brains, but not the 'diffused education without which the ideas of men of genius cannot fructify beyond the limited scope of an individual'.

Accepting Dewar's estimate of 1,500 chemists we may infer that there were, in England, at most, 3×107 graduate chemists; or, limiting ourselves to holders of British degrees, 3×75, or 225. If we limit ourselves still further, to those who held only British degrees, the number becomes 177. We are quite justified in selecting the last number, for, while we are not concerned with foreign graduates, the double-graduate can hardly be considered typical of English educational practice; double-graduates were probably the sons or heirs of factory owners, like the famous Muspratt brothers of 50 years before. It can also be inferred that graduates were less likely to be over-looked by the committee than were non-graduates; and certainly the 'tail' of the chemical industry was unlikely to be rich in graduates. We may

* It can be added that the German dye industries, etc., were now financially and scientifically impregnable. It was the more unfortunate that these great plants could easily be converted into munitions works. That this was done in 1914 goes almost without saying; so that, in the final reckoning, the price was paid not only in trade but also in lives lost.

conclude, therefore, that the chemical industry* at that date employed about 180 to 200 graduate chemists with an upper limit of some 225. The term 'graduate chemists' includes men graduated in English university institutions, or evening classes, and dependent for their livelihood on their employment; that is, employees and not owners, or part-owners.

It cannot be said that industry offered great opportunities to the increasing numbers of young science graduates. What appointments, then, could they expect? Hardly the Civil Service, for the Royal Commission of 1913 [29] showed quite clearly that the higher grades of that body were the preserves of Oxford and Cambridge, and more particularly of the Classical Greats and Classics Tripos schools within those universities. Apart, therefore, from one or two minor and ancillary professions, the one significant vocation left is that of teaching.

From the *Schoolmaster's Yearbook* of 1903, the first year of issue, it appears that, of 3,870 graduate masters whose names were listed, some 510 were scientists; that is, they had taken either Natural Sciences at Oxford or Cambridge or held a B.Sc., M.Sc., or D.Sc., degree of one of the new universities. (The number does not include those who had taken degrees in mathematics.) Thus at least 10 out of every 76 graduate schoolmasters held science degrees. It is of very great interest that less than 1 per cent of these men held foreign degrees; a proportion that compares remarkably with the 48 foreign graduates out of a total of 107 in the chemical industry.

We have already seen that, in 1897, there were some 8,000 graduates engaged in secondary teaching. It will hardly be an exaggeration if it is assumed that there were the same number of graduate secondary teachers in 1902, for, in the five intervening years, there had occurred a marked expansion in secondary education and an equally well-marked trend therein to employ graduate teachers in preference to non-graduates.† A proportion of 10 out of 76 would, in this case, yield a total of over 1,000 graduate science teachers in secondary schools. How many of these can we suppose were chemists? A proportion of

* Including printing, dyeing, acids, metallurgy, explosives, oils, fats, soaps, paints and varnishes, brewing, confectionery, pure chemicals and pharmaceutical products, sugar, starch, glucose, cement, tiles, pottery, aniline dyes, tar, paper pulp, glue, gelatine, size, paraffin, dyewood, tanning.

† *Report of Royal Commission on Secondary Education*, Vol. 1, p. 238; and Vol. V, p. 167.

about 40 per cent would be a very reasonable figure; it can be justified on the ground that chemistry was, and I believe still is, the most popular of all the science subjects, enjoying a marked lead over physics, its nearest rival. Granting this, we would have a probable number of 400 graduate chemists engaged in secondary teaching.

But this does not exhaust the vocational opportunities of the teaching profession. Graduate chemists were also engaged in other fields of teaching. Taking the universities: there were, at that time, approximately fifty graduate chemistry teachers in the grant-aid universities and university colleges. To them we should add the chemists of Oxford and Cambridge universities, the Royal College of Science and the university extension colleges such as Reading and Southampton. We should also take account of the graduate chemists in polytechnic and technical college teaching; that is, in those institutions brought into being by the various Technical Instruction Acts from 1889 onwards. Finally, there are those who were teachers in primary schools and those engaged in what may be described as ancillary educational services: inspectors etc. This second group of graduate teachers could not have numbered less than about 150 and may well have been much larger.

Bearing in mind the assumptions made, we reach a final estimated total of about 550 graduate chemists engaged in the various teaching services. Combining this with the estimated number of chemistry graduates in industry (between 180 and 230) we have, as a total of active, professional chemistry graduates engaged in chemical industry and in teaching some 730 to 780; and of these, as we have seen, 550 or between 70 and 75 per cent were teachers. It is merely a corollary of this to say that the majority of those who were reading for degrees in chemistry at that time were doing so in the anticipation of becoming teachers; and, if this was true of chemistry, it must *a fortiori* have been true of physics and the other 'pure' sciences, for we have seen that the applied physics laboratory was subsequent to the applied chemistry laboratory.

It is evidently desirable to examine in more detail the educational changes at that time; not from the point of view of the educationist but to discover the opportunities and openings which the profession offered to the graduate scientist. Opportunities which, of their nature, depend on the organic relationship that exists between the primary, the secondary school, and the university; a relationship which, in turn, depends on the state of social welfare and the public notions as to the

O.S.E.—8

educational opportunities to be offered to the various grades of society. These latter, of course, are complex questions; too complex by far to be discussed here. Yet it must be remembered that, in the last resort, the development of professional science is governed by the leading ideas of social justice that spring from and, in return, condition the complex web of social relationship.

That very characteristic Victorian institution—the Department of Science and Art—was ultimately replaced by the Board of Education by an Act of 1899. This was followed by a second very important piece of legislation which, in 1902, constituted the Local Education Authorities, with power to organise and harmonise both elementary and secondary education. In 1895 some 128 organised science schools were receiving grants from the Department; in 1901–2 the State, for the first time, aided the secondary schools as such, and in 1902 the organised science schools became Division A schools of the Board. To quote the official record, 'For some years after 1902 the efforts of the State and of the local education authorities were mainly devoted to augmenting the supply of secondary schools' [30]. The regulations of 1907 laid down the principle that the aim of secondary education was not to educate the working class élite but to bring the opportunities of good education within the reach of all. Certainly the direct intervention of the State was followed by a rapid expansion of these secondary schools:

Year	Grant-Aided Secondary Schools
(1895)	(128 Organised Science Schools)
1905	491
1906	600
1907	676
1908	736
1909	804
1910	841
1911	862
1912	885
1913	898
1914	910

That is to say, the State secondary schools had been doubled in ten years, and by 1914 there were seven times as many as there had been science schools twenty years before. Probably a number of these schools had existed before the State intervened, but it is certain that this action meant a great gain in efficiency and in the quality of the education provided. Also there was a strong reaction against the inhumanity of the earlier science teaching; against the cramming and grinding of the older education, and this produced great improvements in the methods of science teaching.

At the same time there was a rapid extension of scholarship awards to carry youngsters from the elementary school to the secondary school. Statistics for the years before 1900 are very difficult to obtain, but it was estimated by the Board of Education that there were some 2,500 scholarships in 1894. The rate of increase here was fairly rapid:

1894	2,500
1900	5,500
1906	23,500
1912	52,583

Figures which speak for themselves. The corresponding increase in the number of State secondary school children was from 31,000 in 1902 to 151,000 in 1912. However, it was estimated that of the 40,000 children annually leaving secondary schools by the latter date, only about 2 per cent were going directly to the university; this was not regarded as satisfactory.

When we turn to elementary schools we find much the same kind of expansion and liberalisation going on. The role of science in elementary education was admirably defined in the regulations for 1904; and there was a great increase in the number of student places in the day training colleges which, as we have seen, were now associated in many cases with universities and university colleges. Thus the available places increased:

1890	3,679
1900	6,011
1905	8,987
1910	12,625
1913	13,098

And many students in university training colleges were taking degrees; a development noticed by the Royal Commission in 1895.*

It is quite obvious that such great changes could not take place

* Roscoe, Vol. III, p. 203.

without affecting very profoundly the production of the trained scientist. At one end of the scale there were increased opportunities for the talented boy to acquire at the new secondary schools a far better education than his father could have enjoyed, and, at the other end, there was evidently going to be a post, a teaching post, for him to fill when he had graduated. In many respects this great reform movement was the long overdue achievement of that which Germany, with her *Gymnasia* and *Realschulen* had accomplished such a long time before. What this meant for the universities can best be described in the words of a contemporary observer, Ramsay Muir. Writing in 1907 Muir [*31*] commented that the great spread of universities over the last fifteen years had brought university education within the reach of thousands to whom it had previously been unattainable and 'Hence has come a remarkable increase in the "natural supply" of teachers adequately trained at their own expense. But this is not the only result. Casting about for students, the new universities perceived, in the primary teaching profession, a vast field waiting to be cultivated; and as the supply of training colleges was quite inadequate to meet the demand for teachers, and the universities or university colleges could cheaply supply this need, they succeeded in obtaining from the government a licence to train, in regular university courses, large numbers of students whose expenses were paid out of public funds on condition that they undertook to devote themselves to the profession of primary teaching. To-day these students form in every British university an appreciable element, and in the new universities a substantial proportion of the total number of students. These students are trained side by side with and (in so far as their non-professional training is concerned) in precisely the same way as other students who intend to become secondary teachers.'

The gist of Muir's analysis is that there were, in the arts and science faculties, two groups of students: the would-be secondary teachers who were self-supporting fee payers and a large group of grant-aided training college students destined for the elementary schools. The former constituted the 'natural supply' of secondary school teachers and, to quote Muir, '. . . fearful of cutting off the natural supply . . . we have laid down that no primary teacher may be recognised as a secondary teacher. We have not been able fully to enforce this rule, but still, it is our rule.' Thus formal sanction was given to the curious belief that the primary teacher should be trained at the State's expense, while the secondary teacher should pay for his own education. This was described by Muir as preposterous; it was the more so since the supply of secondary teachers seemed to be running out, in spite of the augmented universities, and some local authorities were being forced, tacitly, to use primary teachers, or training college students to staff their new secondary schools.

If, judging by present-day experience, we find it strange that large numbers of science graduates should be primary teachers, we should remember that, at a time when other vocational opportunities were few, the officially controlled primary school offered security and a not unreasonable salary. Also it opened the way to either a headship or a lectureship in a teacher training college, both reasonable possibillities for the graduate teacher.

Thus, considering the primary and secondary school together, the great expansion of the educational institutions of the country is obviously directly correlated with the rapid rise in the number of graduands (see Figures 2 and 3). In fact, this is the main implication of Muir's remarks.* Certainly, the Board of Education noticed the large proportion of young teachers in secondary schools in 1907-8; only 28 per cent of the men and 15 per cent of the women were over 40; this was ascribed to the very recent growth of the institutions. By 1912 the Board was complaining of the shortage of secondary teachers, in spite of the fact that in 1907 the regulations for the elementary training colleges were revised, and the training of secondary teachers regulated and subsidised by the Board. The 'Declaration' was amended in that those entering the elementary training colleges were required to declare merely that they would serve for seven years in a State school without limitation to elementary teaching. This met Muir's criticism; for the elementary graduate was now free to enter a secondary school if he wished and if there was a vacancy.

We can see the importance of this development by setting out the figures for the day training college students who took degrees together with the totals of degrees awarded by the grant-aided university colleges. These figures have been derived from the *Statistics of Public Education in England and Wales (1899-1914)* and from the *Returns from Universities and University Colleges in Receipt of the Government Grant . . . (1894-1914)* [32].

* There is evidence to suggest that before the great expansion graduates had had difficulty in getting appointments (see p. 181)

(I)	(II)	(III)	(IV)	(V)	(VI)	(VII)	(VIII) Students with Degrees taking Teaching Diplomas	(IX) Grand Total of Teachers
Year	B.A.s		B.Sc.s		Totals			
1899	91	—	119	—	210	—	—	—
1903	146	—	207	—	353	(98)	—	—
1907	271	(115)	293	(103)	564	(218)	—	282
1908	299	(112)	338	(119)	637	(231)	64	310
1909	312	(138)	312	(108)	624	(246)	79	323
1910	381	(135)	462	(118)	843	(253)	77	324
1911	386	(204)	461	(186)	847	(390)	71	499
1912	438	(197)	464	(162)	902	(359)	109	473
1913	503	(190)	512	(145)	1,015	(335)	114	442
1914	490	(192)	476	(136)	966	(328)	107	—

The final column gives the number of clearly intending teachers graduating in each year. It is made up of the elementary teachers (Column VII) and those graduates who, in the following year, took the teachers' diploma of the universities; it being assumed that the students in Column (VIII) had actually graduated in the previous year.

The very high proportion of teachers is obvious. Out of 3,080 Arts graduates between 1907 and 1914 inclusive, no fewer than 1,283—or 43 per cent—were would-be elementary teachers. There were 3,318 science graduates in the same period, of whom, 1,077, or 33 per cent, were elementary teachers. Combining the totals between 1908 and 1913 and adding the teacher's diploma students we find that, of 5,432 graduates, no fewer than 2,653, or 49 per cent, were clearly intending teachers.

Turning now to the full-time degree students in the arts and science faculties we have the following figures:

Year	Science		Arts		Total	Elementary Teachers
1911	—	3,500	—		3,500	1,459
1912	1,504	(697)	1,866	(747)	3,370	1,444
1913	1,421	(607)	1,772	(756)	3,193	1,363
1914	1,401	(610)	1,805	(716)	3,206	1,326

Here it seems that the proportion of elementary teachers in the science faculty is slightly higher than the proportion in the arts faculty; the percentages of elementary teachers in the former faculty being 46, 42 and 43 for the years 1912, 1913 and 1914 respectively.

It is quite evident that the majority of the degree students at the universities and colleges up to 1914 were intending teachers. It must be remembered that 33 per cent of those who took science degrees had signed a binding agreement to teach for seven years; it can hardly be supposed that those were the only would-be teachers in the universities' science faculties. There must in the nature of the case have been many scholarship holders and fee payers, who intended to become teachers of one sort of another. Muir took it for granted that—up to 1907 at least—the would-be secondary teacher paid his own fees. We should therefore be perfectly justified in putting the percentage of science-degree students who were to become teachers at well over 50 per cent as a preliminary estimate. In any case it is quite obvious that no other single occupation, as for example the chemical industry, could possibly have been a factor in determining the career of such a large percentage of students as 33–46 per cent.

In the year 1913–14 the total number of post-graduate students in the grant-aided colleges who were either studying for higher examinations or doing research in science was 172 in England and 17 in Wales. This is an astonishingly small figure and may be taken as a natural consequence of the career-bias towards secondary and elementary teaching. Even the leading physics laboratory in the country—the Cavendish—could claim only about twenty-five research students in the year just before the war; and this despite the eminent talent that had guided the laboratory. If we take into account science research students at Cambridge, Oxford and Imperial College as well as the university grant colleges, it appears very improbable that there were, in England and Wales, more than three hundred science students doing systematic post-graduate research just before the outbreak of war.

In the years preceding 1914 the rate of increase of full-time students slackened and, in the case of the arts and science faculties the numbers actually fell. This was due to a fall in the number of training college students—a fact which did not escape the attention of the Board of Education.

Year	Training College	Others	Total
1911–12	2,126	5,701	7,827
1912–13	2,070	5,596	7,666
1913–14	1,982	5,774	7,756

These figures relate to full-time students only. The 'others' include, besides arts and science students, medical, engineering, law, agricultural and technical students, and comprised both degree and diploma candidates. A number of the full-time training college students were, of course, working for diplomas and not degrees.

In the meantime the expansion and rationalisation of the State secondary schools was continuing:

Year	Graduate Teachers	Untrained Graduate Teachers	Total Teachers	% Graduate Teachers
1907–08	3,651	..	7,581	48
1908–9	4,278	..	8,436	50
1909–10	4,685	2,568	8,825	51
1910–11	5,057	2,754	9,077	53
1911–12	5,411	2,873	9,126	55
1912–13	5,720	3,057	9,430	59
1913–14	6,076	..	9,810	61

The very large number of untrained graduates calls for comment. Recapitulating briefly; between 33 per cent and 46 per cent of science undergraduates were, as training college students, committed to the teaching profession, and, in addition, there were a few graduates who took university diplomas in education. All these students, when graduated and otherwise qualified, were regarded as trained teachers; but, even so, less than half the secondary graduate teachers were trained, and this implies that, unless the Oxford and Cambridge contribution was disproportionately large, approximately 33 per cent of undergraduates in addition to those classified as 'clearly intending teachers' subsequently entered the teaching profession—probably in the form of Muir's 'natural supply'. Moreover, the rate of increase of untrained graduate teachers was practically equal to the rate of increase of trained teachers. It is therefore reasonable to revise the preliminary estimate of 50 per cent and to suggest that at the very lowest 66 per cent of science undergraduates at universities and colleges were, at that time, destined for the (primary or secondary) teaching profession.

The number of teachers in secondary schools with no experience or with less than one year's experience, is given below, together with the estimated number of these who were graduates:

Year	No. with no experience over 1 year	Number of Graduates
1909	1,830	930
1910	1,914	1,010
1911	1,984	1,090
1912	2,087	1,230
1913	2,185	1,330

These figures give some idea of the demand for graduates that the secondary schools were making at that time. Even allowing for the numbers of Oxford and Cambridge graduates entering the State secondary schools, there were evidently very many who came from the grant-aided universities and colleges. In fact, the number of raw graduates is comparable to the total of arts and science degrees awarded during these years; and this takes no account of the other teaching requirements—elementary schools, technical colleges, etc.

Thus, summarising, quantitative and qualitative evidence has been put forward to support the proposition that at the end of the nineteenth century the majority of science graduates adopted the profession of teaching rather than any other. Similarly it has been shown that the great increase in the number of science graduates for all universities during the years 1895–1914 is accounted for by the extensive re-organisation in the fields of primary and secondary education with the correlative increase in the number of teaching posts available.

TECHNOLOGICAL EDUCATION

With the expansion of secondary and university education and with the measures of social welfare that enabled increasing numbers of young men and women to benefit by a higher education that in former years would have been quite outside their grasp, the technical education movement as such loses much of its importance for us.

In contrast to German practice, technological education in England was spread thinly over the country in numbers of relatively small technical colleges. In the former country it was, as we have seen, concentrated in a small number of excellent polytechnics, well endowed and equipped and of university status (the degree being Dr.Ing.). The evening class continued to be the main mode of instruction in England whereas in Germany it was much less commonly used. But, whatever the formal differences between the practices of the two countries, it

o.s.e.—8*

was generally agreed that, in England, both day and evening classes in higher technology were not in a very satisfactory condition. In the Report for 1908–9 [*33*] the Board of Education maintained that 'The slow growth of these technical institutions is, however, in the main to be ascribed to the small demand in this country for the services of young men well trained in the theoretical side of industrial operations and in the sciences underlying them. There still exists among the generality of employers a strong preference for the man trained from an early age in the works, and a prejudice against the so-called "college-trained" man.'* They added, a little ingenuously, that harm had been done by colleges that claimed that their training dispensed with the need for practical experience. It was hoped that the new secondary education would, in time, remedy this state of affairs; but, in the meanwhile, 'It is to be deplored that there are several schools in which the well-qualified staffs and the excellent equipment practically stand idle in the day-time through lack of students'.† In every yearly Report up to the outbreak of war the same complaint was made by the Board of Education.

A possible explanation for this inertia was put forward by Professor Raphael Meldola who had had experience both of industrial science and of education. Meldola was disappointed that the 23 polytechnics in and around London and the 110 in the provinces had not turned out to be centres of research; he believed that 'the danger is the general tendency in this country to ram the whole scheme of education into one mould utterly regardless of the fact that the requirements of, let us say, an engineer are quite different from those of a chemist'. The teachers were heavily called on and had little time for research although many of them were competent men. Also objectionable was the 'baneful system of teaching subjects in "classes" so that a syllabus qualifying for some particular examinations may be gone through in a certain time. It is quite unnecessary to point out here that individual originality or the spirit of research can never exist in such an atmosphere.' The student can never assimilate the subject as a living principle and the teacher, however original and zealous at the outset sooner or later deteriorates. 'That is one of the reasons why the polytechnic movement has produced such a small effect upon the chemical industry' [*34*]. A further injurious practice is 'the statistical standard by which the success of these institutions is chiefly, if not absolutely judged'. In Meldola's view: 'A school of science which is not also a centre of

* p. 90.   † p. 171.

research is scarcely worth the cost of maintenance'. If this is true, then there was some substance in his criticism for, while there were 300 posts for teachers of chemistry in these 130 institutions, the output of original research from them was not representative of the powers of 300 men; only about twelve were doing research, and that desultorily. His plea was for creative and scientific technologists and to this end he made a demand for 'the general recognition of research as an educational discipline . . .'.

A common hindrance to the growth of scientific and technological education was the burden of fees that the student had to bear. Fees had often been discussed during the nineteenth century debates on higher education but, curiously enough, never it seems as a matter of some urgency. Gregory had, as we saw (p. 64 above) pointed out in 1842 that because of an ample State subsidy, the fees at Giessen were only £6 14s od. a year; much less than any Scottish university could afford to demand. And it appears that this differential persisted throughout the century. Thus, in 1902, the Registrar of King's College, London, wrote to *Nature* to correct an exaggerated report of the cost of tuition in chemistry at that college, but conceding that '. . . unfortunately there can be no dispute as to the great difference between the fees charged in Germany and in England' [*35*]. Universities tend to be rather coy about the fees they charge and there are obviously many factors that make it difficult to compare the full costs of university education in different countries. Nevertheless other evidence exists to confirm the view that university education was much cheaper in Germany than in England at the beginning of this century [*36*].

The ruling philosophy in England throughout the nineteenth century appears to have been that an individual who wanted higher education must be prepared to pay for it, even in colleges that were maintained in part by the State. Accordingly, fees from students continued to be the greatest single source of revenue in the grant-aided colleges and universities up to 1914, when it amounted to about 32 per cent of the whole.* It is not surprising therefore that, as Jowett had observed, the lower middle classes could not afford to send their children to London or provincial university colleges. In spite of the efforts of men like Playfair and Lockyer the scholarship ladder was very inadequate even at the end of the century: a mere two thousand or so from primary to secondary school and about the same number

---

* In Germany the State subsidy was by far the biggest item in the universities' revenue accounts.

from secondary school to university, including all faculties. The scales were therefore heavily weighted against talent from the lower classes; from, that is, the great bulk of the population. One member of Lockyer's deputation to meet Balfour, a Mr. Bell, M.P., had pointed to the importance of reducing university fees and the deputation had formally requested 'the enfranchisement of proven ability in all classes'. But real progress in this direction was not made until later in the century.

THE WAR YEARS AND AFTER

The first clear indications that the *fin-de-siècle* decline in British science was being halted was, as we have remarked, the work of Larmor, Stoney and J. J. Thomson on the electron and of William Ramsay and Morris Travers on the inert gases. That there was some substance behind the revival was confirmed by the achievements of the small but able schools of Rutherford, Soddy and Aston. The numbers were, of course, small compared with those of the great European schools of science at that time.*

From the industrial sector, too, there was evidence to suggest that British engineering industries were beginning to recover some of their initiative after the long years of torpor [37] and, indeed, that more attention was being paid to applied science. Even the aniline dye industry, such as it was in England, began to look up a little [38] and it was reported that while 21 firms had accepted graduate scientists as chemists, metallurgists, geologists, etc., between 1906 and 1910, 40 firms did so between 1911 and 1914. This improvement, in view of the educational advances, was only to be expected.

To speculate on what might have happened had there been no war would be unprofitable; yet the rapidly rising output of science graduates from all universities does suggest that applied science in England might have made comparatively big advances during the years given over to war and the succeeding distress. But, in August 1914, all the main developments described above came to an abrupt end. The thread of continuity was broken, and we have no means of

* But there is some evidence to suggest that the relative lead of the German universities was beginning to diminish. The growth of strident nationalism was hardly in accordance with the high ideals—and practice—of nineteenth-century German universities. Furthermore, there was an increase of State interference in university affairs to such an extent that Ostwald resigned his Chair in protest.

telling what the future development of 'pure' and applied science would have been had war not broken out.

As is shown by the dramatic fall in the number of students taking Part II of the Cambridge Natural Sciences Tripos (see Figure 2) there was, in effect, no national scientific policy.* The pursuit of science and the training of scientists came to an abrupt halt in 1914; in this respect 1939-45 was completely different, as the same diagram shows. At the same time, and for obvious reasons, all German institutions, including universities, fell into immediate disrepute. For example, some medical men, wishing to discredit certain aspects of the Haldane Commission Report, complained, in October 1914, of the 'Attempted Germanisation of London University'. What a radical change! Most, if not all, reformers had been in the habit of comparing German universities very favourably with British ones; this attitude was no longer possible, and it would not be unfair to say that German universities have never since 1914 been regarded in this country with the admiration they once commanded.

Whatever the orientation of public opinion the shortages and deficiencies consequent on war with Germany could not be overlooked. There was an immediate shortage of dyestuffs, for Britain manufactured only one-tenth of its requirements. The country had also depended on Germany for magnetos, for drugs and countless pharmaceutical preparations, for tungsten used in steel making and for zinc. Less than twenty-four kinds of optical glass were manufactured in Britain compared with over one hundred made in Germany. Before the war Britain had been in the habit of obtaining some of the chief materials for explosives from Germany and, actually, at one time, had no means of producing acetone. It needed Sir William Ramsay to point out to the (presumably classicist) civil servants that such apparently innocent substances as cotton and lard were potential explosives. Lacking a policy, there was no procedure for taking account of science, and in this respect the country was little more advanced than it had been at the time of Charles II when the Royal Society had been founded. It was not really excusable when we recall the repeated warnings of Col. Strange, the Devonshire Commission, Lockyer and many others, to say nothing of Playfair's remarkable prophecies. It was officially admitted that 'the necessity for the central control of

---

* Although figures are hardly necessary to prove the point; the most promising young physicist of his generation, H. G. J. Moseley, was killed in action at Gallipoli.

our machinery for war had been obvious for centuries, but the essential unity of the knowledge which supports both the military and industrial efforts of the country was not generally understood until the present war revealed it in so many directions as to bring it home to all' (1916).

One of the first acts of the government was to create in 1914 the British Dyestuffs Corporation, with a total capital of £3 million, of which sum £1½ million was found by the Treasury, and at the same time £100,000 was made available for research. On 23 July, 1915, the coalition government issued a White Paper outlining their 'Scheme for Organisation and Development of Scientific and Industrial Research'. On 28 July, 1915, a Committee of the Privy Council for Scientific and Industrial Research was formed with an advisory council* with the object of instituting specific researches, establishing special institutions for applied science and endowing research studentships and fellowships. In their first report [39] they outlined some factors which they believed had contributed to the slow development of applied science in England. One of the most conspicuous was the small, individualist scale of English industries; the manufacturer had felt that research on the small scale he could afford, would be at best a doubtful proposition and, in any case, the joint stock banks were reluctant to finance such enterprises. On the other hand many urgent problems were of such a nature that only the largest organisations could afford to carry them through; could, that is, afford to be really scientific. The Council concluded that the current need was for the establishment of research institutions to fill the gap between university science and industry. This necessitated a supply of trained researchers for, while it is true that demand creates supply, it is also true that supply creates demand. Before the war the output of university graduates had been far too small to meet even a moderate expansion in demand for research. In fact the annual output of first and second class Honours students was running at only 550 a year in mathematics, science and technology and few of these had had any research training. We had not yet, the Committee said, learned how to make use of mediocre ability, and without scientific rank and file it would be impossible to staff the scientific research laboratories of the future.

The development, from this time onwards, was fairly rapid. Related

* Comprising Sir W. S. McCormick, Lord Rayleigh, O.M., Sir G. T. Beilby, W. Duddell, F.R.S., Prof. J. A. McClelland, Sir C. A. Parsons, Raphael Meldola and Richard Threlfall.

to the D.S.I.R.* for that was the body with which we have been dealing, there grew up a number of Research Associations: the British Photographic Research Association,† the British Scientific Instruments Research Association, the Motor and Allied Industries, Iron, Glass, etc., etc.‡ Besides this, the D.S.I.R. had taken over the National Physical Laboratory and created a number of Research Boards; the Fuel, Building, Forest Products, Radio Research Board and several others.

While this was being planned, a number of public men had formed a 'Committee on the Neglect of Science'. They met in conference on 3 May, 1916, with Lord Rayleigh in the Chair [*41*], and all united in demanding a wider diffusion of science among the populace. Significantly a number of classicists supported this and even urged that the stranglehold of the classics on the Higher Civil Service examinations should be broken and science should be included. In the opinion of J. J. Thomson, university and college scholarships should be awarded for all round knowledge and 'not as at present for highly specialised knowledge of one subject. If this were done, neither the literary specialist, ignorant of science, nor the scientific specialist, ignorant of literature, would have a chance—this, I think, would be a good thing—for it would stop that specialisation before boys leave school which is doing so much harm.'

Among those present at this conference was H. G. Wells and Thomson's words may well have been in his mind when he attended the eleventh annual meeting of the British Science Guild a year later [*42*]. H. A. L. Fisher had said that England had failed 'so far to find a form of scientific instruction which appealed to the imagination and the interest of the general mass of school children who are not destined for what I may call a specifically scientific career'. Wells, speaking as 'an old schoolmaster' took the argument on to a much more general plane. His words are worth repeating: '. . . you cannot get a more general interest in science at the present time since you have no class of persons to get the general mass of people in touch with contemporary scientific work; because scientific men are, generally speaking, scientific specialists, ignorant of philosophy and literature, and without any

* Department of Scientific and Industrial Research, now the Science Research Council, or S.R.C. [*40*].

† The first; licensed by the Board of Trade on 15 May, 1918.

‡ There are now over forty of the Associations; sponsored jointly by the government and by industry.

bridge between them and the man of ordinary education. (Laughter.) No. Don't laugh! These are serious things. The ordinary man cannot reach over to the scientific specialist and the scientific specialist cannot reach over to the ordinary man. There is a gap in our public mentality at the present time. It is by no means a comic gap.' Wells went on to blame the universities for the 'Greek shibboleth', for, 'It splits and divides our national consciousness by setting up a barrier that cuts science off from philosophy and history. We cannot get along with our scientific men cut off from the general thought of the community, and the general ideas of the community cut off by a devotion to the dead languages from the stimulus of living science.' He demanded that philosophy and history be freed from the 'Greek shibboleth'; for, 'Until you do that your man of science will still be an unphilosophical specialist and get as much respect as he does today, and your literary and political men will be unscientific, unprogressive and unenterprising, full of conceit about their "broader outlook", and secretly scornful of science'. This, according to Wells, was the 'fundamental disease'.

A syllabus which, had it been adopted, would have satisfied Wells' requirements was suggested by the Prime Minister's Committee on Science in Education [43] under the chairmanship of J. J. Thomson. They recommended for sixteen- to eighteen-year-olds, a course in science to include the history of science and development of scientific ideas, the lives and works of scientists and a study of the relations between science and industry.

One of the main difficulties of the advancement of science was also touched on by this committee. As they remarked: 'The greatest advances in pure science are often the outcome of investigations which, until they are justified by success, appear fantastic and unpromising, and meet with little approval from orthodox scientific opinion, and it is too often a long time before any tangible results are obtained; for this reason they are not of a kind which could be expected from workers in a great institution supported by public funds'. The difficulty is that a man would, assuming he was in process of making a great advance, have little to show for his salary. It is extremely difficult to see how, in modern society, such an objection can be met. Naturally, the committee made no attempt to tackle it, nor, indeed, were they expected to. In the past such enterprises had either implied great wealth, for example, the Earl of Rosse and the wealthy endowers of the Great American telescopes—Yerkes, Lick, etc.; or a remarkable degree of self-sacrifice on the part of the investigator, as, for example was demonstrated by J. P. Joule and John Dalton. To-day we are, perhaps, at once better and worse off. On the one hand the costliness of certain branches of scientific research—notably nuclear physics—has made them matters of State endowment, for they are utterly beyond any private purse. It is quite probable that the great nuclear machines, and radio telescopes in use at our universities, greatly exceed in cost all the money ever spent on science by the State from the earliest times up to 1910. On the other hand there are a number of liberal foundations which in some measure do the work of the old patrons of science; and this is very important for the interest of the State in science is a matter of immediate national requirements rather than disinterested love of knowledge.

With the end of the war we have reached the very important modern period; a period worthy of study on its own account. The apparatus of the modern scientific world has now been assembled, and all the familiar outlines are present for, at the end of the war, the physical industries—as we should term them in distinction from the chemical industries—immediately began to found their own research laboratories. It should be pointed out that an incentive towards this was the development of the thermionic valve; a scientific consequence of the work of Crookes, Fleming, Langmuir and others, which was to some extent the physical equivalent of the discovery of the aniline dyes in chemistry so many years before.

REFERENCES

[*1*] C. F. Carter and B. R. Williams, *Industry and Technical Progress,* (Oxford University Press 1957), p. 108 et seq.

[*2*] H. J. Habbakuk, *American and British Technology in the Nineteenth Century* (Cambridge University Press, 1962).

[*3*] L. T. C. Rolt, *Tools for the Job*, pp. 214–16 and Robert S. Woodbury, *History of the Lathe to 1850*, p. 74 et seq.

[*4*] Extensive details of these firms may be found in John H. Dunning, *American Investment in British Manufacturing Industry* (Allen and Unwin, London 1958).

[*5*] D. S. L. Cardwell, *From Watt to Clausius, the Rise of Thermodynamics in the Early Industrial Age.*

[*6*] *The Royal Society Index, 1800–1900* (Cambridge University Press, 1912) iii.

[7] Arnold Toynbee, *Lectures on the Industrial Revolution* (London 1884).

[8] J. T. Merz, *A History of European Thought in the Nineteenth Century* (4 vols., London 1896–1914).

[9] Arthur J. Marder, *From the Dreadnought to Scapa Flow* (Oxford University Press, 1961) i, pp. 30–1

[10] J. N. Lockyer, *Education and National Progress*, p. 118.

[11] Fabian Ware, *Educational Foundations of Trade and Industry* (London 1901).

[12] Arthur Shadwell, *Industrial Efficiency: a Comparative Study of Industrial Life in England, Germany and America* (London 1905) p. 637.

[13] *The Times*, 18 January, 1901.

[14] Ibid, 6 November, 1902.

[15] E. H. Starling, 'The Pressing Need for More Universities', *The Nineteenth Century* (June, 1901) il.

[16] *Nature* (30 May, 1901) lxiv, p. 109.

[17] Presidential address to the Royal Society, 1902.

[18] Sir Philip Magnus, 'Some Economic Aspects of the Application of Science to Industry', Oxford Summer School, 21 August, 1903. Printed in *Educational Aims and Efforts*.

[19] *The Times*, 28 June, 1903.

[20] *Reports of Royal Commissioners on university education in London, 1910–1913* (Haldane Commission).

[21] Sir Arthur Schuster *Universities Review* (1905) i.

[22] *Universities Review* (1905) i, Pt. 1 and Pt. 4.

[23] P. J. Hartog, 'Universities, Schools and Examinations', *Universities Review* (1905) i.

[24] P. J. Hartog, 'Examinations in Their Bearing on National Efficiency', *J.R.S.A.*, 3 and 10 February, 1911.

[25] J. N. Lockyer, 'The National Need for the State Endowment of Universities'. Statement as President of the B.A., (1904). See also *Nature* (1 January, 1903) lxvii, p. 193 and (21 July, 1904) lxx, p. 271.

[26] *The Times*, 29 January, 1902.

[27] H. W. Macrosty, *The Trust Movement in British Industry* (London 1907) Appendix 4.

[28] *Report of the 1902 Meeting of the B.A. at Belfast*, pp. 15–18, 97–8.

[29] *Report of the Royal Commission on the Civil Service*, Cd. 7338, (1914), 4th Report, p. 42.

[30] *Report of the Board of Education for the Year, 1911–1912.*

[31] Ramsay Muir, 'The Supply of Teachers', *Universities Review* (1907) iv.

[32] *Statistics of Public Education in England*, Cd. 568 (1899–1900);

Cd. 1139 (1900–01); Cd. 1476 (1901–2); Cd. 2782 (1903–4–5);
Cd. 3886 (1905–6–7); Cd. 4288 (1906–7–8); Cd. 4885 (1907–8);
Cd. 5355 (1908–9); Cd. 5843 (1909–10); Cd. 6338 (1910–11);
Cd. 6934 (1911–12); Cd. 7674 (1912–13) and (1913–14).
*Returns from Universities and University Colleges in Receipt of the
Government Grant*, Cd. 7459 (1894); Cd. 8165 (1896); Cd. 8984
(1898); Cd. 9410 (1899); Cd. 331 (1900); Cd. 845 (1901);
Cd. 1510 (1902); Cd. 1888 (1903); Cd. 2366 (1904–5); Cd. 2789
(1906); Cd. 3885 (1907); Cd. 4875 (1908); Cd. 5246 (1909);
Cd. 5872 (1910); Cd. 7008 (1911–12); Cd. 7614–15 (1912–13);
Cd. 8137–8 (1913–14).

[*33*] *Reports of the Board of Education for the year(s)*, *1908–9*, p. 90;
1909–10, p. 118; *1910–11*, p. 138; *1911–12*, p. 95; *1912–13*, and
1913–14, p. 111.

[*34*] Raphael Meldola, *Fifth Annual Report of the British Science Guild*
(7 April, 1911).

[*35*] W. Smith, *Nature* (11 December, 1902) lxvii, p. 127.

[*36*] See, for example, *Handbook of British, Continental and Canadian
Universities, with Special Mention of the Courses Open to Women*,
edited by Isabel Maddison (New York 1899) pp. 64, 87–100.

[*37*] David S. Landes, 'Technological Change and Development in
Western Europe, 1750–1914', in *Cambridge Economic History of
Europe*, edited by H. J. Habbakuk and M. Postan (Cambridge
University Press, 1965) vi, Pt. 1, p. 565 et seq.

[*38*] W. P. Dreaper, 'The Research Chemist in the Works', Institute
of Chemistry lecture (London, 1913).

[*39*] *Report of the Privy Council for Scientific and Industrial Research for the
Year 1915–16* (22 August, 1916). Ibid, *1916–17*; *1917–18*;
1918–19, etc. *Scheme for the Organisation and Development of
Scientific and Industrial Research*, Cd. 8005 (1915). See also Ian
Varcoe, 'Scientists, Government and Organised Research in
Great Britain 1914–16: The Early History of the D.S.I.R.',
Minerva (April, 1970) viii, pp. 192–216.

[*40*] *Minerva* (April, 1970) viii, pp. 304–7.

[*41*] *Report of Conference at Burlington House* (Committee on the Neglect
of Science) 3 May, 1916 (London 1916).

[*42*] H. G. Wells, *Eleventh Annual Report of the British Science Guild*
(June, 1917).

[*43*] *Report of the Prime Minister's Committee on Natural Science in
Education* (London 1919 and 1927).

CHAPTER EIGHT

The Professional
Society

Up to the outbreak of the Great War the development of technology
and of applied science in England, judged by the supply of skilled
labour and by the provisions made for training and education, had
been carried forward by a sequence of responses to public events. The
Great Exhibition of 1851, or rather the lessons allegedly learned from
it, led directly to the South Kensington enterprise, to the science
section of the Department of Science and Art and also prepared the
way for later developments in technical training. The alarms and
revelations of the Paris Exhibition of 1867 gave rise to the great
technical education movement of the seventies and eighties, which
was more concerned, generally speaking, with engineering than with
applied science. It can thus be distinguished from the later movement
in the nineties and early years of the twentieth century which was
stimulated by the well-known German achievements in applied science
and which helped to produce increased State aid for universities as well
as the establishment of Imperial College and the Manchester College
of Technology.* During each of these times of panic a great deal was
accomplished and much more hoped for—we recall the persistent
attempts to found an industrial university—but after a time the sense
of urgency relaxed and matters were allowed to drift until the next
alarm; the process was, therefore, one of fits and starts. There was in
fact, no settled policy for orderly, evolutionary development; there
were only responses to awkward and unavoidable situations.

In the absence of a policy—indeed, of a coherent philosophy of
higher education—it is hardly surprising that applied science did not

* The old Manchester Mechanics' Institute had evolved during the
nineteenth century into a Municipal School of Technology. When, in 1903,
Owens' College was granted its Charter as an independent university one of
the conditions laid down was that it should establish a faculty of technology
in the Municipal School of Technology. This was done and the faculty
subsequently expanded to take over the whole of the college, which is now
known as the University of Manchester Institute of Science and Technology
(U.M.I.S.T.).

relatively speaking, play a major part in the Great War. This was not the case in the Second World War, of course; science and technology were immensely important between 1939 and 1945. Now the problems that were set and solved in the fields of physics and chemistry are fascinating in themselves and in the talent and ingenuity evoked to resolve them. But the social problems set and solved—the mobilisation of the abilities of a large number of 'pure' and applied scientists—seem to have aroused little interest. This is all the more strange when it is recalled that sixty or more years ago such achievements even on the relative scale, would have been quite impossible. It is therefore with professional scientists, whether 'pure' or applied, that we must now deal.

It is mainly with scientists of the second and third rank that we shall be concerned, but before we discuss them we must consider, if only briefly, those of the first rank. As we have seen, the machinery for high grade, specialised instruction had been assembled by the universities and colleges by about 1881. We should naturally expect that this would, sooner or later, affect even scientists of the first rank. It is indisputable that the Royal Society represents the pinnacle of natural science in Britain, election to a Fellowship being regarded as the highest distinction for a scientist. That rationalisation has modified the composition of the Society in the direction of increased specialisation and correlative professionalism is clearly indicated by a brief analysis of the vocations of Fellows in 1881, 1914, 1953 and 1967.

	1881	% increase	1914	% increase	1953	% increase	1967
Distinguished laymen	54		38		8		11
Sailors	13		6		2		0
Soldiers	26		6		3		2
Applied Scientists	62	27%	79	70%	134	23%	163
Academic Scientists	134	116%	289	20%	348	36%	475
Medical men*	55		11		6		2
Clergymen*	14		4		0		0
Others*	120		40		46		17

* Clergymen and medical men in 1881, 1914 and medical men in 1953, 1967 are those who were or are unassociated with any academic or applied science foundation. Many of the 'others' in 1881 were distinguished amateurs: Darwin, Joule, Spottiswoode (P.R.S.), etc. Today there are no amateurs.

These figures do not require detailed explanation. The university take-over of science, to which we referred on page 191, is quite apparent. It is now virtually complete; many of those classified as 'applied scientists' are in posts that are effectively interchangeable with university appointments—for example, with the Atomic Energy Authority, the Medical and Agricultural Research Councils, the National Institute for Medical Research and the Marine Biological Association.* At least twenty-three but not more than fifty applied scientists are employed in productive industry either by private enterprise firms of by State Corporations.

There has occurred, during the present century, a virtual exclusion of all men not engaged in physical or biological science in a professional capacity. The excision of the amateur element has been almost surgical in its neatness. And not only the amateurs but also those engaged in sciences other than the physical and biological have been removed from the lists. In 1881 and in 1914 there were about six Fellows who were distinguished for their contributions to social science; to-day there are about two and both of these gained their Fellowships before they transferred their interests from natural to social science.

The percentage increase in academic Fellows between 1881 and 1914 is over four times greater than the percentage increase in applied scientists. This is not surprising when we reflect that the period was conspicuous for the foundation of new universities and university colleges, for increased State aid for education and for an increasing student population. In short it was a time of educational revolution and higher education was relatively more important than applied science. On the other hand the period between 1914 and 1953 was marked by the converse process: the proportion of applied scientists increased much more than did that of academic scientists. This was almost certainly due to the tremendous expansion of applied science that may have begun just before the outbreak of the Great War in 1914 but which was undeniably expedited by that war and by the following World War of 1939–45.

Any comprehensive social, economic or political study of science in contemporary Britain must take account of the Royal Society; it

---

* In all probability the Ideal Type Fellow of the Royal Society to-day would have been educated at one of the older British universities, would have spent some time at an American university (but not a German or other European one) and would be, or have been a teacher in a university.

would, in fact, be disrespectful to do otherwise. We have already suggested that the general reforms initiated during the nineteenth century may have unintentionally imposed disabilities on potential scientists among the less privileged classes; it would be interesting to know what changes have occurred in consequence of the more drastic and extensive reforms of the last seventy years. One surprising change has been that the relative increase in number of Fellows over the last eighty years (about 40 per cent) is much less than the relative increase in the number of practising and qualified scientists in the country. If the admissions to Fellowship had kept pace with the numbers of scientists there would now be several thousand Fellows. Whether this would be desirable or not is a matter that is worthy of debate in an educated democracy that is also anxious to ensure that science flourishes.*

Let us, however, turn now to scientists of the second and third rank, for they comprise the bulk of those engaged in applied science and in the different grades of teaching. Statistics of scientists and technologists in education, in government service and in industry have become in recent years much more comprehensive [1]. These indicate that, in 1965 no less than 46,563 scientists with university degrees of equivalent qualifications were employed in industry or by research associations, government or local authorities. If we restrict ourselves to chemists and physicists we find that 21,860 chemists and 8,846 physicists were employed. The same statistics indicate that for every four scientists employed in management, sales and similar functions five are employed in research and development. Thus we may conclude that 12,000 chemists and 5,000 physicists are engaged in what we may broadly classify as industrial applied science.

The statistics show that in the same year some 43,756 scientists were employed in education, of which total some 9,722 were chemists and

* Equity suggests that all qualified scientists who have contributed original papers to recognised journals should be eligible for election to Fellowships. But the touchstone of equality of treatment may not be applicable in this case. The real objection to élite bodies is not that they are élite but that experience shows that they often oppose necessary changes and that they sometimes make serious mistakes by excluding really creative men and important new ideas. Again, why should the number of Fellowships be restricted to six or seven hundred? There may be good reasons why a management unit or an army battalion should be about this size; but do such reasons apply to the leading scientific society in Britain and, probably, the world?

7,490 were physicists. The implication of these figures is plain: industry has replaced teaching as the main occupation of natural scientists. If we transfer the totals of university scientists—2,653 chemists and 1,822 physicists—from 'education' to 'research and development' then we may say that in both these two basic sciences, research and development is now more important than education.* In the case of the biological sciences and of mathematics however the situation is reversed: education is the main field of employment and relatively few are employed in industry and specifically on research and development.

Clearly there has been great progress since Dewar's address to the British Association in 1902 and the two hundred or so graduate chemists in the chemical industries of that time. A large corps of professional applied scientists and scientific technologists has been created; science is being used systematically and on a national scale to minister to the needs of society, to extend mastery over the natural environment and—most important—to devise entirely new processes, materials and instruments for social use. What is so surprising is the extreme newness of the activity; it is at most some seventy or eighty years old. A group of men, numbering a few hundred at the beginning of this century, has been expanded to an unprecedented extent and used to create a series of entirely new technologies as well as to transform and rejuvenate the old established ones.

In his evidence before the Devonshire Commission, Edward Frankland estimated that, in 1845, about twenty persons a year received instruction in practical chemistry in England; none, of course, received instruction in practical physics. If his estimate was substantially correct, and we have no reasons to doubt it, there cannot possibly have been more than a few hundred scientists of the second and third rank acting in a professional capacity in England a century ago. Such an estimate will include the staffs of observatories and museums, surveyors of the ordnance and geological offices, a small number connected with the defence forces (F. Abel, for example) and the scattered few in industry. We can illustrate this by comparing the social structure of science a century ago with a slender tower, almost as broad at the top as it was at the lower levels and ground floor. It was inferred that it was the absence of ground floor development that was the cause of the pessimism of those who, starting with Davy and Babbage, lamented the decline of English science. Playfair recognised it and Huxley, as

* That is to say 14,600 chemists and 6,800 physicists will be engaged wholly in research; 7,000 chemists and 5,600 physicists wholly in education.

we have seen, summed it up accurately when he spoke of the absence of 'rank and file'. These two were by no means the only ones who understood the deficiency; and by 1916 it was officially admitted that England had still not yet learned to use 'mediocre talent'. To-day, with the relatively tiny Royal Society and many thousands of second and third rank scientists the social structure of natural science can be correctly compared with a pyramid, many times broader at the base than at the top.

Having outlined the establishment of professional science let us now consider in more detail than before, the relationships between applied science and industrial innovation. Francis Bacon, we remember, had recognised two distinct types of invention: one, which we have called empirical invention, is substantially independent of scientific knowledge, while the other is essentially science-based and can only be effected if appropriate knowledge is available. The importance of the second type justified Bacon in urging the advancement of science. It also influenced his many subsequent disciples, including the supporters of the great mechanics' institutes movement. But it seems that only now, in the modern applied science laboratory has the procedure been institutionalised.

In general terms an applied science laboratory may, we suppose, serve two purposes. It may be a place where scientific inventions are made or perfected; where, that is, scientifically trained people using technical data and knowing the relevant laws of nature can invent such things as new electronic devices or new chemical products, which are, of their nature, implied in existing scientific knowledge and whose invention therefore does not involve extending the frontiers of knowledge. On the other hand the laboratory can be a place where researches, methodologically indistinguishable from 'pure' science, are carried out into the properties of materials or into the nature of processes in which the organisation may be, or may become interested. Of course, the laboratory may well be concerned with both these activities. It is, in the absence of much more detailed information, impossible to say what proportion of those whom we classify as applied scientists are scientific inventors and what proportion are researchers. As J. R. Ravetz has remarked, a training as a scientist is quite appropriate for these kinds of activities for 'a scientist once adjusted to the new sorts of problem will work as well as someone trained as a technologist' [2].

Besides applied scientists of the types we have just mentioned, innovation may also involve technologists who we may define as people with both scientific knowledge and practical experience. Technologists can design efficient products or processes in accordance with theoretical principles and at the same time be capable of operating within known practical limits. In this sense the technologist is not an innovator, a creator of original ideas or inventions, although of course someone trained as a technologist may well be qualified for employment as an applied scientist if he so desires. Finally there are those technologists who apply the Smeatonian procedure for systematic evolutionary improvement (see p. 14 above). By this procedure the motor car for example, has been improved out of all recognition in the last sixty years although its basic design—or invention—has not changed appreciably in that time.

The modes of innovation are clearly many and in our classification we have had greatly to simplify them. The factors affecting successful innovation, whether highly scientific or crudely empirical, are also many and complex. Nevertheless this has not prevented many people sharing the cosy illusion that we (the British in this instance) make the original inventions or discoveries that foreigners then exploit for their own profit. It is easy to sustain this prejudice, common among the European nations, since most people are more familiar with the work done by their own nationals than with the, often vital, contributions made by foreigners.

A great deal of light was thrown on the factors affecting scientific innovation in Britain by the investigations of the Science and Industry Committee which was set up by the British Association, the Royal Society of Arts and the Nuffield Foundation in 1952 and which carried out an extensive survey of British firms during the following six years. One finding of the Committee* that is relevant to our discussions is that firms that are successful in applying new scientific ideas tend to be 'open' ones in that there is a good deal of free communication between the various persons whose different scientific and technological skills are necessary to launch an innovation. Such a firm must have scientifically trained staff not only in the laboratory but in the production and sales departments and in management generally. If a research laboratory is instituted in default of these conditions—perhaps for reasons of fashion, or prestige—it may well fail and its scientists become frustrated. It was found too that where scientists are represented on the board of directors the firm is more likely to have had a successful

* The author worked for the Science and Industry Committee between 1954 and 1956.

record in innovation than if scientists are not on the board. But this may well be an effect as much as a cause: an efficient firm will recruit scientists and it will use them effectively so that in the fullness of time those scientists who have administrative gifts will rise to become directors.

Among the salient characteristics of scientifically progressive firms that the Science and Industry Committee noticed [3] were that they tended to take in a wide range of high quality scientific journals, similar to those taken in by a large university science department; that they were surprisingly willing to share their knowledge with other firms and to encourage their scientists to contribute to journals, conferences, etc.; that they looked for new ideas from outside the firm, and indeed often from outside their country; that they accorded science and technology very high status and that they knew how to attract and keep talented people. These characteristics suggest that a firm that becomes scientifically progressive must take on some of the aspects of a university science department. This, on reflection, is not too surprising since both are concerned with scientific information and with the forwarding of scientific knowledge, so that what is good for the one must, within limits, be good for the other.*

One might even argue from this that because the applied scientist is concerned with the investigation of laws of nature he warrants the title of scientist and the term applied scientist should be used—as it often is—to describe the work of the technologist. This, however, is unacceptable; the researches of the applied scientist are guided not by purely scientific considerations, but by the requirements of industry. The hallmark of the scientist is his absolute freedom of inquiry; he may wander at will from one field of knowledge to another as the nature of his researches and the impulses of his mind lead him. This freedom of research and its enormous importance for the advancement of science has been demonstrated many times in the history of science; to imagine that science can dispense with it and be unaffected thereby is surely a most dangerous illusion. But this does not mean that the applied scientist and the technologist are to be regarded as truncated scientists; their primary duty and loyalty are not to abstract knowledge but to the economic welfare of their firm or industry. Some students

* It would seem, in retrospect, that the hierarchical and authoritarian ideas of British education and, indeed, of British industry must have made the evolution of such firms very difficult if not impossible during the nineteenth century.

of scientific industry suggest that there may be a clash of interest between management, with its economic orientation, and scientists whose loyalties lie in part at least with the external profession of science. But I doubt if too much should be made of this. If many industrial scientists are concerned about their professional status this may be for the shrewd reason that anything that enhances their status also enhances their earning power. No doubt there are 'pure' scientists *manqué* in industry but in their cases the solution is simply that they should find alternative and more congenial employment with a research association or council or in a university.

Technology, judged on the basis of its scientific content, has, of course, existed for as long as the sciences with which it is concerned; and, although it merges imperceptibly into skilled craftsmanship it is, in one sense, antithetical to the latter. The aim of the great technical education movements was, above all, to increase the scientific element in technology at the expense of empirical, rule-of-thumb, or traditional skills. The technical colleges in England, the polytechnics in Germany and Switzerland were established for this very purpose, and with their expansion went the corresponding developments of scientific technology and a reduction in the domination of old-fashioned craftsmanship, made necessary by the increasing number of industries founded on the results of contemporary scientific research. This raises a very important question. Is the practice of applied science merely a development or refinement of technology? In other words, did the class of technologists evolve, in due course, the class of applied scientists?

So far as such a question admits of an answer, it must be in the negative. The evidence shows that the first professional applied scientists were usually trained as 'pure' scientists and that the first industrial research laboratories—in Germany—were derived rather more from university practice—from Giessen and Heidelberg—than from the workshop or plant. It is a straightforward deduction from this that, before industrial scientific research can be established on a notable scale, a number of conditions must be fulfilled. Respectively, they are: (1) A number of those concerned with the running of industry must have an adequate knowledge of science. (2) It must have been shown that scientific research can be effectively carried out when deliberately restricted to certain problems, (i.e. can be 'guided') and that such researches can produce results that can be usefully applied within a reasonable time. (3) It must be appreciated that researches of this nature do not require first rate talent; quite moderate abilities suffice, and (4), there must be an adequate supply of professional scientists available. The nature and success of the Ph.D. system put (2) and (3) beyond dispute as far as Germany was concerned while the general education system of that country ensured that (1) and (4) were satisfied.

It is reasonable to claim that, not until the closing years of the nineteenth century did any of these conditions come near to being satisfied in England. Indeed, their fulfilment necessitated the abandonment or profound modification of some well established English traditions and ideals: notably the ideal of the self-taught, self-made leader of industry, the long traditions of higher education and, not least, of learning and science. The meetings of the British Association ninety or a hundred years ago were patronised by a heterogeneous collection of parson-naturalists, university professors, philosophically inclined medical men and lawyers, engineers and amateurs, wealthy and not so wealthy. The absence of a group of professional applied scientists meant that the relationships between science and technology were both loose and indirect, despite individual instances to the contrary. In fact, in the past, scientific revolutions have been accomplished, as in the seventeenth century, without associated technical changes while technical revolutions have taken place without effecting immediate changes in contemporary science: as was the case during the industrial revolution. To-day, science and technology are very closely linked; as the development of nuclear energy, for example, shows.

Sociologists and historians have pointed out that military discipline had to be introduced before the use of firearms became effective in warfare. There is surely a marked parallel here with the development of applied science? We note that the applied scientist is, like the competent soldier, a thoroughly trained man and, again like the soldier, he is subject to discipline; a discipline that is, however, economic and social rather than physical. The training of the professional scientist is generally in a 'pure' science and is carried at least to the point at which he is competent to carry out his own researches; preferably research should have formed part of his training. The element of (social) discipline is provided by the fact that he depends on the practice of science for his livelihood.

The beginning of true applied science has been dated, with some precision, as occurring between 1858 and 1862, and in the first instance it appeared in the German dyeing industry. Surprisingly, for as we have seen, there was every incentive for England to render herself independent of foreign dyestuffs as she had the world's largest textile industry. An Englishman had made the actual discovery, the raw materials were both abundant and cheap in England. The capital was available and there were many commercially enterprising Englishmen at that time. It may be argued that foreign 'protection' and English patent laws were deterrents, but such excuses are special pleading. What is undeniably the case is that at the critical time there was no class of professional, highly trained chemists in England; no large group of men generally recognised as capable of carrying out research in chemistry and, at the same time, expecting to earn a livelihood by doing so. Similarly there were very few chemists who had the knowledge and skills for the other functions that were vitally necessary in a highly scientific industry: technical sales, production and management generally.

The aniline dye industry was in one respect, very unusual. It sprang solely from an advance in science: in organic chemistry. The traditional animal and vegetable dyes were replaced by new, scientifically compounded, substances. Science was, we may say, the master key; it was not invoked merely to refine and improve methods or materials that were either immemorial or the result of rule-of-thumb development. At the same time it became clear that systematic research would yield more and even better dyestuffs from coal-tar; and it was in this respect that Germany excelled. She had to import the raw coal-tar from England, but, to process it, she had her large corps of professional research chemists on which her industrialists could draw. This was a point foreseen by Playfair: that the development of transportation would soon reduce local advantages in raw materials and skill would become a determining factor. Moreover, once the success of scientific research allied to industry had been demonstrated beyond doubt the rest followed quite naturally and we cannot be surprised by Germany's subsequent successes in the electrical engineering industries, etc. (This is no more than an assertion of the well-known fact that the location and development of particular industries is governed by the available supply of skilled labour; it being assumed that industrial scientists can be regarded as skilled labour.)

We can now gather the threads together and conclude that the emergence of applied science depends on the inherent opportunities of the situation (see p. 10). The internal development of science produced a material of great social utility and, at the same time, gave promise of even greater treasures along the same line of research. The issue, that is, was clear cut; a scientific challenge was posed. To respond to a challenge of this nature a society must have the services of a class of professional scientists. In the case of the 'physical' industries the development of the thermionic valve, with its enormous possibilities for research, invention and correlative utility, presented a challenge in some ways analogous to that of the aniline dyes. But, whatever the field in which the challenge occurs, it must be in a precise and easily grasped form; it would have been theoretically possible at any time, during the period examined, to have established research laboratories in connection with the mining, metallurgical and engineering industries, and undoubtedly these would have led to great benefits. However, quite apart from the absence of professional scientists, the need for science in these industries was diffuse—it could not be brought home to the industrialists in a concise, inescapable form.

So far we have avoided the major issue. If there is no applied science how can a class of professional scientists be created? And if no class can be created how can there ever be applied science? This looks like a vicious circle and such, indeed, was the case over most of the nineteenth century in England. But it overlooks the point that has been developed above: the professional scientist can also be a teacher, he need not be limited to industry. In fact, it is clear that, in principle, before you can have a class of professional scientists you must create the necessary educational machinery. This needs no proof whatsoever. But there is the additional factor: the very act of creating a suitable educational machinery also creates, *sui generis*, the professional scientist. This you cannot avoid doing unless by law, statute or custom you deliberately exclude science from the syllabuses. But then your educational system would be inefficient and retrograde as it was in England for so long.

We may pause for a moment to consider the significance of the Mathematics Tripos in the history of professional science. This justly famous course of training in mathematics and theoretical physics spans the whole period of this work and during that time it underwent less modification than any other course of study; it implied, from the beginning, a highly specialised education and to this general form all other courses of study have gradually approximated. It is very natural to ask, therefore, why it was that the Tripos did not produce in England a class of professional scientists.

Quite apart from the technical reason that mathematics and theoretical physics are of more limited applicability than chemistry and experimental physics, two distinct answers can be given to this question. Firstly, the great majority of earlier candidates were young men of means and, as such, did not constitute the best human material for what is, from a certain point of view, the banausic art of applied science; moreover, tempting careers in law, etc., were open to the young Wranglers. Secondly, when the opportunity occurred the Tripos course did, in fact, produce the professional scientist. It would be supererogatory to show that most public school mathematics masters were Tripos men; and when the development of university colleges and provincial universities began in earnest in the seventies and eighties Cambridge found herself in the position of natural supplier of teachers and professors of mathematics and physics for those institutions.

The importance of teaching posts for the development of professional science had been foreseen by Lyon Playfair, James Hole, Henry Latham and T. H. Huxley before such posts were made available, and therefore before professional science began. It follows that this is not a question of wisdom *ex post facto*. We can see how, in practice, it works when we consider the case of a well-organised and smooth-running system of secondary and trade or technical schools together with effective universities, all of which require competent science teachers. Of course a good system of primary schools is also necessary and, at the same time, a very liberal scholarship ladder to accommodate the talented. Young men and women, educated at primary and secondary schools are vocationally trained at university to be returned to the former as teachers; they therefore perpetuate the principle. But some of these students—a small minority perhaps—will have talent above the ordinary and will want to do research at the university. A proportion of these will normally be retained as university teachers; but as supply generally exceeds demand in this case, those who do not become university teachers must either revert to school teaching, or abandon science, or become applied industrial scientists. From this point of view therefore the industrial scientist is to be regarded as an internal product of the educational system. Of course it is not asserted that applied science is in any way 'inferior' as a vocation to university or school teaching; there are, to-day, those who undergo research training with the express intention of becoming industrial scientists.

The origin of the German achievement has been traced, as we have seen, to the development of the Giessen laboratory from 1825 onwards. But beyond that was the educational policy laid down by von Humboldt in the first decade of the nineteenth century. In any case the Giessen venture began long before German industrial expansion, before German industries were in any way capitalist* and before even the Zollverein was accomplished. From a comparatively early date the integrated German educational system—the primary schools, the State secondary schools: the *Gymnasia*, the *Progymnasia* and the *Realgymnasia* that so aroused the admiration of Matthew Arnold and others, the trade schools and the polytechnics and universities all offered opportunities for the science teacher at different levels. We saw that Conrad ascribed the sharp increase in numbers of students in the philosophy faculty partly to the extension of this educational machinery; and Friedrich Paulsen later commented that the philosophy faculty which in the eighteenth century provided preliminary training for the other faculties (divinity, law and medicine) added, in the course of the nineteenth century, the function of providing teacher training. Later on, according to Paulsen, the faculty became conspicuous for scientific research and the training of advanced teachers. Such a system, which sent up young men well-prepared to the universities, could hardly fail to produce the professional scientist. There would always be those who would resolve on a career of learning, and those men would form the man-power potential of the new group. The ultimate deciding factor, therefore, must have been the educational machinery; the necessity of staffing the universities, polytechnics and schools. Indeed it is easy to believe that, in the 1850s, the number of professional scientists in German schools and universities must have exceeded those in German industries. Certainly the German educational revolution preceded the applied science revolution by quite a few years; and it would be difficult to believe that this was carried out in the interests of, then uninvented, applied science. On the contrary, the German academic can be very fairly charged with some degree of intellectual snobbery; Liebig, as we saw, thought that science in Germany was valued for its own sake and not, as in England, for what it produced. I gave a brief and, I hope, cogent reason for my belief that he was wrong with regard to England. One can accuse Liebig of ignorance of his own countrymen, but it is less easy to dismiss the professors' semi-contemptuous phrase—'*Brotstudien*', and certainly the German universities strongly resisted the introduction of technology into the

* Sir John Clapham.

O.S.E.—9

university syllabus.* They were successful in this but it did not prevent them resisting, unsuccessfully this time, the raising of the polytechnics to university status—and this was long after Oxford, Cambridge and all other British universities had engineering faculties or departments.

It is apparent that the education system in England did not offer the same opportunities as did that of Germany. The universities adapted for the training of the English gentleman, were largely closed to research and the advancement of learning, as the nature of the degree system and examinations shows, to say nothing of the explicit statements of educationists. In the society of the time this did not seem anomalous (save to enthusiastic reformers and admirers of the German system) for science was, as we have seen, prosecuted with vigour and success by a brilliant group of semi-amateurs. The great public schools, in organic relationship with the universities, were naturally governed in accordance with the same educational aim, and the possibilities of science masterships were few until the 'reform'. There remained the new university colleges, but these were bedevilled by poverty and by the external degree system which made systematic study difficult to achieve; moreover even here the degree was liberal both in form and content.

For the rest little need be said. The mechanics' institutes relied, for many years, on part-time or voluntary lecturers and the standard must have been low and erratic. Also they were in decline due to bad primary education and other causes adduced above. Of the primary schools nothing need be said.

The first effective step was the foundation of the various university science examinations. But the pace of development was governed by the quality of the students coming up from school—and testimony is fairly complete that most were ill-prepared—and on the jobs available on graduation. As we saw from the Devonshire Commission and from the minutes of London University, it was only when the schools began to employ science masters that graduate science became a profession;

---

* The term 'pure' science does not seem to have been used in English writings before the middle of the last century. It is likely that it, and the idea underlying it, was introduced into British scientific circles by the many chemists and other scientists who studied in Germany during the nineteenth century and thus learned to admire German science and scientific organisation. The notion of 'pure' science does not seem to accord with British traditions which were empirical and even utilitarian. In Germany on the other hand it marked the division between the universities and polytechnics.

and we saw, too, that the reorganisation of the South Kensington colleges as a Normal School resulted in a flood of applicants that was greater than the available accommodation.

Such evidence is qualitative rather than quantitative. But there is one other aspect which must not be forgotten. The scientist must have had a practical training if he is to deserve the name. However, systematic practical classes did not become common until the later 1860s and practical examinations began later still. That is, many of the 'science' men turned out before that date were merely book-learned; at best they might know about science, at worst they would know only how to pass science examinations. Thus, while this practice prevailed, the universities could not produce scientists, still less could the sterile State examination system—witness the boy who 'passed' in 19 subjects.

The group of academics that met at the Freemasons' Tavern in 1872 wanted to create a class of research men, which class, they said, did not exist at that time. It can hardly be supposed that these men did not know what they were talking about; that, in fact, there was such a class and they were unaware of its existence. Even if this is held to be inconclusive, that the professional scientist did not then exist was proved beyond question by the fact that the Department of Science and Art was forced to employ officers of the Royal Engineers as school inspectors, for the reason that they were, at that time, the only men in England with a professional training in science. From the sixties to the nineties England was unable to muster as many as sixty professional scientists to serve as part-time inspectors of her technical schools.

With the gradual reform of the secondary schools, the passing of the various Technical Instruction Acts and the founding of the university colleges, the opportunities for science teaching were greatly expanded. But the biggest factors seem to have been the elementary training colleges, and, from 1902 onwards, the new State secondary schools. We have seen that the rapid expansion in numbers of degree students before the war was linked first to primary and then to secondary educational developments rather than to any other factor. This meant the production of graduates who would be financially dependent on science teaching, and this implies the professional scientist. At the same time the old universities were being rapidly liberalised from within, which meant, of course, the admission of potentially professional students from poor homes. The numbers of successful candidates for Part II of the Natural Science Tripos, which had fallen steadily from 1881 to 1900 now began to rise steeply (see Figure 2).

To pursue this matter further would be to go into the general sociology of education, and that would lead us away from the point. We have established, and I do not think it can be doubted, that the professional scientist is, in the first instance, the product of the educational system; to a much less extent is he the product of industrial practice and economic organisation. Or, to put the claim at its lowest, the applied science revolution cannot possibly be understood without reference to the reform of the educational system. To-day there are frequent complaints of the shortage of teachers of science in schools, coupled with observations that industry has tempted too many graduates away from teaching. If these complaints are substantially true, then it looks as though scientific industry is devouring its own *alma mater* and this would be very unfortunate. This forms part of the sociology of science to-day, and in the absence of further knowledge, any comment would be no better than speculation. None the less it would be very interesting and is important to know how the streams of natural science are divided among the various professions and what factors govern that distribution.

The old professions of law, medicine, etc., rested to a considerable extent on a personal relationship between the expert and the client; service was rendered to the community through the individual [4]. The professional scientist, on the other hand, must render his service to the individual through the community; an individualist relationship between the scientist and a client or employer, is unthinkable. In this respect the professional scientist is more akin to the civil servant or the army officer than to the doctor or lawyer; indeed he transcends even the former pair in the impersonal nature of his services. Therefore it would follow that the level of professional science is a function of measures of public welfare, or of collective social action. Especially true is this of the initial stages, for the lengthy and expensive education of the natural scientist necessitates State aid on a substantial scale; that has been the universal experience of every country that has been able to develop applied science. The denial of State aid during the crucial period 1850–80 was the final reason why applied science was later in making its appearance in England than in Germany.

The theories of self-help and of individualism were given a full and fair trial in nineteenth-century England; whatever achievements are to their credit—and they are many—they proved, when applied to science and education, quite incapable of producing the professional scientist. As the century progressed this fact was recognised by an increasing number of people; from the 1870s onward the doubters were in the majority, at least as far as science was concerned (see above especially the Devonshire and Samuelson Commissions).

I am not here concerned with generalisation as to such large scale social institutions as 'capitalism', etc. Such generalisations would not, in any case, teach us much, nor would they take us very far. In this context we should note that the free nature of American and Swiss political and economic institutions did not prevent them from liberally endowing their educational and scientific foundations while the more rigid and differentiated class structure of Germany was likewise no obstacle to scientific progress in that country. But we are not concerned to discover ultimate causes, still less to pass moral judgment on men who, if they failed to see what the minority saw, were not therefore to be condemned. Here, we are concerned with the development of a certain small group of men and a certain mode of social action together with the conditions that governed the evolution of the group and its functions. It needs to be emphasised, I think, that applied science is itself an invention and it is an invention that can only be effected in a certain type of society at a certain stage of development. The first scientist or industrialist to suggest the permanent employment of a research scientist, or group of scientists, was the person who invented applied science.

Even if we assume that industrial requirements on their own can break down an iron curtain of Latin and Greek, buttressed by class privilege and underpinned by the Established Church, we can hardly suppose that they could create a syllabus of studies and researches which is not understood by the educational authorities of the time. In whatever way industrial requirements express themselves, whether through the explicit demands of employers or otherwise, they cannot be wiser and see further than the received notions and ideas of those concerned with education. The point is that the utility of research, of laboratory practice, of applied science in the modern manner, is not at all obvious in the first instance. The properly trained scientist has been educated in a way and to a standard that would either be a luxury or useless when judged by the criteria of the day-to-day needs of industry; his employment cannot be justified after a week or a month or perhaps even a year's work.* What, the industrialist might have asked, is

* Cases were quoted to early investigators (e.g. the 1884 Commissioners) of German applied scientists whose work had been fruitless for as long as two years. But it was added that such men justified themselves in the long run.

meant by 'scientific research'? The question is a difficult one to answer; it exercises the wits of philosophers to-day. It cannot be expected therefore that an industrialist could understand the value of science *a priori*—he must see it in action before he can incorporate it in his industrial activities. If there is no way in which it can be brought to his attention and if there are no professional scientists available, it follows that he cannot utilise applied science.

I have already produced material evidence to suggest what I have argued in principle above—that until the universities were producing the specialist, industrial demand could not make itself felt—did not, in fact, exist—and young men could not enter industrial research in large numbers. This is a reversal of that theory which explains professional scientific training by reference to industrial demand—it is an assertion of the opposite. The reverse of the professional coin is, of course, a form of specialism; but specialism, as I have shown, arose as a socially selected consequence of university examinations, of Honours Schools and Triposes, and in such a manner that demand for professional applied scientists could not possibly have played any part in the process. Further, I went on to show that the first real professional scientists to be produced in any significant numbers were not would-be industrial researchers, but intending teachers; and it was only when those conditions were fulfilled that an industrial demand could, at last, arise. Industrial demand therefore played no more part in the specialisation of the various science degrees than it did in the specialisation of the Classics Tripos or the London B.A. Nor can industrial demand be credited (or blamed) for the form which specialisation has, generally speaking, taken.

Apart from universities, and university scholarship examinations, the specialisation of studies would be most congenial to schools. The division of labour in the scholastic world was of long standing, and a good 'Honours' man was an asset to the school. But once the educational revolution was well under way and professional scientists were being trained in increasing numbers, did it follow that the adoption of applied science by industry was inevitably and smoothly effected? To answer such a question, framed in the most general terms, would involve a study of the diffusion of ideas among the community and a wide study of the requirements and responses of different industries. The factors favouring the widespread adoption of science are, firstly, the diffusion of scientific culture through improved education; and, secondly, the incorporation of science courses in the various kinds of technical education. On the other hand, we have recorded many denunciations of the English manufacturer for his failure to appreciate science—and although many of these denunciations were unfair and even absurd (men like Mather, Samuelson, Swire Smith, Mundella, etc., were liberal and progressive employers and were well aware of the importance of science), it may be that for various reasons the English manufacturer was somewhat conservative in outlook, setting a higher value on practice than theory and experiment and reluctant to believe that anyone could teach him how to run even a part of his business. More material, possibly, was the small scale on which many British industries were organised, for, as H. A. L. Fisher put it, '. . . we are an old country of old and small traditional business . . .'.* Beneath these disincentives there may possibly have been additional psychological factors: a distaste for work in industry and factory on the part of men who might hold the intellectual qualifications for industrial research.

Whatever the shortcomings of the manufacturer—and these are conjectural—the marked deficiencies of the educational system must always be remembered in substantial mitigation. Up to 1902 this was quite obvious; as Mark Pattison had lamented in 1872. 'The manufacturing and commercial interests of the country have outgrown us . . . they no longer regard us . . . they do not think we have got anything worth having. . . .' It is difficult to deny that the manufacturers were not without reason on their side.

Even after the great educational reforms the very small number of post-graduate science students (172) at the grant-aided colleges in 1914 does not argue much devotion to research. As long ago as 1884, the Samuelson Commission had stressed the value of research in industry and in training (see p. 133); Dewar had stated that many of our examination scientists would be of little use in a research laboratory (1902); the Technical Education Board Sub-Committee had urged post-graduate research training and, more recently, J. J. Thomson's Committee and the Barlow Report [5] made exactly the same recommendations; the latter stating explicitly that three years' undergraduate study do not make a scientist. Therefore it is reasonable to suppose that even after the educational reforms were achieved up to 1914, the development of applied science may well have been hindered by the unsatisfactory position of research training as opposed to the written examination system then widely practised. The large number of foreign graduates in industry tends to support this view; it would seem that

* *Eleventh Annual Report*, British Science Guild (1917).

the scientific industrialist placed more faith on the research examination (German) degree than on the written examination (English) degree.

Let us clarify this point. The action of written examinations is, we may say, collective rather than individual. Applied to a class of science students, some of whom will later abandon science, some of whom are intending teachers and some prospective research workers, etc., the final examination may well stimulate the maximum study by the greatest number. It is quite possible that, in a vocationally ill-assorted class, there are many who are greatly benefited by this stimulus; but with regard to the would-be professional scientist the case is doubtful. We have seen that, up to 1914, most leading scientists did not approve of written examinations as the final arbiters of scientific education; not without reason, for surely it is obvious that research should form a substantial part of the training of the prospective research worker? Under these circumstances and from the point of view of the research scientist the value of exclusive written examination can rightly be called in question. Yet wherever the truth lies in a debate between written examination and research degrees, it is, at the very least, probable that had the Victorian industrialist enjoyed the benefit of a supply of trained professional scientists, even men without research training, the emergence of applied science in England would have occurred much sooner than it did and might very well have pre-dated the German achievement. Had it not done so, even then, the industrialist would certainly have deserved the strictures heaped upon him. But this is merely hypothetical.

To summarise the position reached, let us invoke (in imagination) a manufacturer of (say) 1880. Let us suppose that he is a progressively minded man, a supporter of the technical education movement and of the new local university college. This is quite justifiable, for we have seen that there were many such as he. He will favour extended education for all classes and may even have good ideas on secondary education. But if he is asked why he does not use science—research—and scientists in his enterprise he may well reply along the following lines: 'The suggestions that scientists be employed in industry is absurd; as well ask Mme Schumann to teach my daughters to play the pianoforte. A man of science cannot be constrained to follow any prescribed path; he cannot produce discovery to order neither is it desirable that he should be expected to do so. He must, and all experience bears this out, be quite free to go where he will and research in whatever direction his genius prompts. Also we know that, although great benefits flow from science, it may take many years before such discoveries are of use, and even so we cannot predict just what use they will be. No industry could possibly afford such an enterprise even if men of science were prepared to serve it'. Scientists, for our manufacturer, mean men like Dr. Joule, Mr. Darwin, the late Dr. Faraday.. .. And if it were pointed out that there were science graduates in the country—there were at most 1,000 by that time—he might be pardoned for taking a somewhat sceptical view of their potentialities. He could not afford to employ one on the off chance that he might be another Faraday, and, if one pointed out what was happening (on a very small scale) in Germany, he would reply that that was all right but he had little time for mere book-learning without experience that transcended the textbooks.

Whatever the industrialists and scientists of the past achieved or failed to achieve, we must remember that we are, ourselves, in a situation similar to that which often confronted them. The present national interest in applied science and technical education is stimulated, as before, by the threats of foreign competition; and many of the views now held, and the arguments being put forward, are, it will be agreed, very similar to those circulating during previous 'technical education movements'.

THE SPECIALISED SOCIETY

The growth of professionalism was one of the remarkable social changes of the nineteenth century. Not only did the long-established professions develop a number of subordinate and ancillary professions but wholly new and distinct ones were founded. As the century progressed the casual, sometimes non-existent, training requirements and the individualistic conditions of entry and practice gave place to carefully prescribed training together with rationalised conditions of entry and defined norms of conduct. The rise of professionalism was, perhaps, inevitable under the circumstances of developing political democracy on the one hand and the rapid advance of mechanical and scientific industry on the other. But, while it would be absurd to ascribe this radical change to the advances of 'pure' and applied science, it would be equally unreal to suppose that science, as a social activity, could in any way avoid the ever-widening net of professionalism.

There are substantial areas of science where such factors as increased cost of research and great complexity of theoretical knowledge do not apply and cannot, therefore, satisfactorily explain the virtual disappearance of the amateur scientist. Nor can recent social levelling tendencies be held accountable, for lack of money has, in the past, proved no deterrent to scientists. Professionalism itself must therefore be invoked to complete the explanation of the decline in amateurism. The mechanism whereby the professional, publicly recognised and approved, must supplant the amateur requires no detailed explanation; it is enough to point out that even those whose means could guarantee them complete intellectual freedom are socially oriented towards admission to the professional circle, and to have little difficulty in accepting the norms of the profession.

Generally speaking the achievement of professional science represents a great advance in social organisation. But certain aspects of specialisation, in one form or another an inevitable corollary of professionalism, are less evidently desirable. Specialisation, as it has developed since about 1850, is frequently characterised by an education that from the age of eighteen or so onwards, is limited to one branch of science, being increasingly directed to narrower sectors; by a continued vocational structure that admits and canalises recognised specialists only and by a society that expects, when it does not demand, that specialists shall continue in their prescribed specialism.

There are, it appears, two stages of specialisation, the educational followed by the vocational. Both merit discussion and, as usual, there are two sides to both questions. Let us consider the educational problem first. As we have shown, leading thinkers in all generations, including the present, have criticised 'undue' specialisation on the grounds that it does not accord with the ideals of a liberal education and that it must result in the fragmentation of culture—we recall H. G. Wells' denunciation of the culture 'gap' some fifty years ago. To combat this, notable attempts have been made recently, particularly by the newer universities, to develop non-specialised syllabuses and to encourage the selection of diverse subjects for degree courses.

The difficulty here seems to be that it is hardly possible to obtain a worthwhile understanding of a particular science without making an intensive study of it. Aristotle was aware of this fact* and it forms the

---

* 'Lack of experience diminishes our powers of taking a comprehensive view of the admitted facts. Hence those who dwell in intimate association with nature and its phenomena grow more and more able to formulate, as the foundations of their theories, principles such as to admit of a wide and coherent development; while those whose devotion to abstract discussions

ground of the argument that research is a most powerful method of scientific education. In a few words, there is no short cut; no easy and painless way to the acquisition of knowledge. It follows that the attempt to combine non-collateral studies runs the risks either of superficiality on the one hand or of overloaded syllabuses on the other. And yet a highly specialised training, if carried over into vocational practice, must have the effect of stratifying the world of learning, dividing one subject from another and thereby making the cross-fertilisation of the sciences more difficult. A community of scientists whose individual disciplines are so specialised that they can hardly communicate with one another evidently does not represent the best arrangement that society can make for the advancement of knowledge. Considerations of this sort were clearly in the mind of Friedrich Paulsen who, writing at a time when the German universities were riding the flood tide of success and prestige, could still complain of over-specialisation and demand a broader approach: 'No scientific study can prosper in isolation. Every science is indissolubly related to others: they presuppose each other as auxiliaries' [6].

In principle, as we have remarked before (p. 148 above) the progress of the sciences is towards unification; more facts and laws of limited application are comprehended in fewer and more general theories so that the whole becomes increasingly systematised and intelligible. It used to be said that the early nineteenth century was the last time that an educated man would hope to understand all the sciences. But this now seems to be an illusion: it is only valid if we assume that the student need bother only about those aspects of the sciences that were later found to be of permanent value. A little history soon convinces us that in all probability he would have 'wasted' a great deal of time trying to comprehend ideas and theories that were subsequently refuted and are now forgotten. We may, indeed, adapt a famous 'law' by observing that in every generation human knowledge always expands to fill the human mind. Specialisation, then, can hardly be entirely ascribed to the growth of knowledge.

We have suggested that the growth and refinement of specialisation are inherent in the nature of the educational system, from primary school to university. Certainly, the peak of specialisation seems to have

has rendered unobservant of the facts are too ready to dogmatise on the basis of a few observations.'

Aristotle, *De generatione et corruptione*, Book 1, 2, 316a, 5 been reached by 1939; at a time, that is, when educational considerations were paramount. With the subsequent growth of post-graduate studies and the rise of scientific research, rather than teaching, as a career for science graduates, we should expect to find that specialisation, at least at the post-graduate level, is being ameliorated. The development of inter-disciplinary studies, such as molecular biology, would seem to confirm this. Nevertheless, it would probably be unwise to assume that there is a general breaking down of artificial barriers and that the problems of specialised education have been solved.

This leads us to a brief consideration of the examination system, the present instrument of specialisation. It will be remembered that among the reasons for the adoption of examinations were the need to devise a means of selection for college office, the need to stimulate wealthy young men to do some systematic study and the desirability of having a demonstrably equitable method of ranking students in order or merit. They were, that is, elaborated at a time when the universities did not conceive it their duty to advance knowledge but rather to provide a liberal education. The system has now been taken over generally, although it is undeniable that the functions of the university and the organisation of science have changed radically. It is a *non sequitur* to suppose that a system designed to encourage liberal education in the nineteenth century must be the best way to train professional scholars and researchers to-day. Lyon Playfair's taunt about 'faithful disciples' still rings uncomfortably true. The Germans, we remember, sought to mitigate this endemic university complaint by encouraging students to work at more than one university in the course of their undergraduate careers. No doubt the obstacles to achieving a flexible and efficient system are formidable, but we are no worse off in this respect than our predecessors. As Huggins remarked, some seventy years ago, university reform has been necessary, is necessary and always will be necessary.

Talent, still more genius, has the essential quality that it is unpredictable. An original thinker is one who, by definition, finds unity where none was previously suspected and to do this he may well have to cross the frontiers of convenience that we erect between the sciences. If, for a long time, a man is effectively discouraged from seeking knowledge in other than prescribed ways, if he must choose either this discipline or that, there is some danger that a creative talent may be frustrated—we remember Hartog's penetrating comment about the 'sensitive feelers of the mind'. The maxim that the cobbler must stick to his last is not necessarily applicable to science.

Two simple but related examples will help to explain this point. The basic science of thermodynamics was established by two men with very different backgrounds. The first, Sadi Carnot, was a young French army officer, the son of an exiled republican, who took time off to study in depth and at leisure the performances of the new steam engines that were engaging everyone's attention at the beginning of the nineteenth century. The result of several years study was a short book that was almost totally disregarded but which was, after the author's death, recognised as a work of genius and the foundation document of a new science. The second man was James Prescott Joule who, as the son of a prosperous Salford businessman, was left sufficiently well off to be able to devote himself entirely to scientific research without having to bother about earning a living. Starting from a study of the battery and the electric motor Joule was led to a series of researches as a result of which he realised that there is a common conversion measure between mechanical energy on the one hand and heat on the other; a measure that is universal and independent of the particular materials or components involved. He was even able to evaluate this conversion factor, the mechanical equivalent of heat as it is called, at first roughly and thereafter with increasing accuracy. For years Joule's work was ignored until at last the tide began to turn and people started to listen to him.

The important common factors that Carnot and Joule shared were that they were both, strictly speaking, devotees of science, that they both met with blank incredulity for many years and that they both had to swim against the stream of recognised institutional science: the sort of science that is applauded by academics and expounded in textbooks. This was the price they had to pay for revolutionising science. It is admittedly possible that a large number of specialised and professional scientists could have done their work, filling in the picture with a detailed mosaic rather than with the bold brush strokes of Carnot, Joule and a few others. But it is by no means certain that this would have happened and it follows as a reasonable proposition that if science is to continue to prosper, psychological and philosophical freedom are as essential as financial and intellectual freedom.

The problems of arranging our institutions to allow for the advance of science must be complex. A component of any satisfactory solution should be, I suggest, the establishment of a critical history and sociology of science. In this context 'critical' may be misunderstood for it is a word that is often misused to-day. Adapting Todhunter's comment we can define a worthwhile criticism as a product of intelligence; it is an original creation and not merely the repetition of a slogan learned from a fashionable political philosopher, past or present. A combination of history and sociology is desirable because without the sociology the history may become little more than a pedantic recapitulation of earlier publications; without the history the sociology will lack any appreciation of the different forms that science can take in different societies at different times and will thus have no valid yardstick with which to assess modern scientific activities. This modest programme is not, perhaps, too much to ask for: after all, government and industry together now spend something like five hundred million pounds a year on scientific research. It is reasonable to ask that some provision be made for the study of the origins and development of science, of its present organisation, institutions and practices and, so far as we can judge, of its future prospects—in the underdeveloped as well as in the developed countries.

A scientific education that made provision for preliminary general culture followed by specialised instruction in the chosen subject and research training together with suitably designed courses in the history and sociology of modern science would go some way to meet the criticisms that have been made by many writers. But it would be only a partial solution. It is hardly less desirable to inculcate an awareness of the full implications of freedom of research. When we consider the careers of men like Darwin, Dalton and Joule, reflecting in particular on their magisterial indifference to the outward trappings of scientific status* and their willingness to spend years in the wilderness, developing and refining their ideas, we can only wonder whether science is still the same activity that it used to be. Such detachment is almost impossible to-day, for the pressures to publish, to win recognition and promotion, to serve on committees and to negotiate with grant-awarding bodies are almost irresistible. But the problems of scale and numbers which have brought all this about—a world Huxley never dreamed of—are not solved by being ignored. If scientific education is to be stimulated by examination requirements and guided by scholastic textbooks, asserting the 'facts' of science but ignoring the

* In those days a man could know, and be known by, most of his scientific peers so that status symbols were unnecessary. This is hardly the case in the much more populous scientific world to-day.

essence—how science is actually carried out, what its powers and limitations are—then the result may be that intelligent and able young people will abandon science for more congenial and stimulating disciplines. As J. R. Ravetz has so well remarked, the consequences of leaving modern, industrialised science to mediocrities may be tragic indeed [7].

It is not suggested that a reasoned liberalisation of scientific education and of subsequent vocational opportunity would notably increase the number of first-class scientists, nor is it claimed that the result would be the immediate solution of specific and intractible problems. All that is asserted is that young scientists should be made aware of the traditions that they should inherit and of the place of their disciplines in the scheme of things, intellectual and social. Furthermore, every allowance must be made for the occasional emergence of the unorthodox thinker, bearing in mind that unorthodox ideas have often proved to be the most fruitful. In a sense the advance of science is marginal in that great advances often occur where and when they are least expected, at any rate by the great majority of people. It is also salutary to remember that there are fashions in science as in every other social activity. Thus nuclear physics was the fashionable science yesterday: its practical importance and philosophical interest made its position in some ways analogous to that of geology 150 years ago. To-day the most interesting science is, perhaps, molecular biology. But no one can predict what will be the significant and fashionable sciences in the years to come.

The argument is, then, that it is of paramount importance to ensure that the social environment of science is the most favourable for its continued development that we can achieve. As Ortega y Gasset put it, we cannot reasonably believe that so long as we provide the money there must be science. We must take account of those things—so far as we understand them—that move men to great achievements in spite of years of frustration and neglect. This means that the world of science must be an open society. It is not disputed that a degree of specialisation is inevitable in the nature of the case. Indeed, the specialised society is now a condition of social advance in respect of the conquest of deficiencies, material shortages, diseases, and the extension of mastery over nature. The problem, it seems to me, is to reconcile such desirable activities with the full development of the potentialities of the individual. There are two aspects to this: the ethical, for we can assert that the development in freedom of all the potentialities of the individual is a constituent of the good; and the utilitarian, for it is only when the individual can so develop that the full material benefits of science, 'pure' and applied, can be realised in society.

REFERENCES

[1] See the *Statistics of Science and Technology* (Department of Education and Science and Ministry of Technology, 1965, 1967). And *Report of the 1965 Triennial Manpower Survey of Engineers, Technologists, Scientists and Technical Supporting Staff*, Cmnd. 3103 (1966).
[2] Private communication.
[3] C. F. Carter and B. R. Williams, *Industry and Technical Progress*, p. 178 et seq.
[4] T. H. Marshall, *Citizenship and Social Class* (Cambridge University Press, 1950) pp. 128–55.
[5] *Report on Scientific Manpower* (Barlow) Cd. 6824 (1946)
[6] Friedrich Paulsen, *German Universities and University Study* (London 1908) p. 322.
[7] J. R. Ravetz, *Scientific Knowledge and its Social Problems*.

Note
Other important papers, besides those mentioned above, have been published in recent years by J. Ben-David, Walter F. Cannon and R. M. McLeod in various historical and sociological journals.

Reference should also be made to *Science and Society*, edited by Norman Kaplan (Rand McNally, Chicago 1965) and to the journal *Science Studies*.

Additional
References

Books, pamphlets, etc. used but not referred to in the text. The more important ones are marked with an asterisk.

[1] *Sir H. E. Roscoe, *The life and experiences of Sir Henry Enfield Roscoe,* (London 1960).
[2] *Sir P. Magnus, *Industrial Education* (London 1888).
[3] *Annual Reports of the National Association for the Promotion of Technical Education* (1888–1907).
[4] *"Industrial Value of Technical Education: the Opinions of Practical Men' (N.A.P.T.E.), *The Contemporary Review* (May, 1889).
[5] *T. H. Huxley, 'Science and Education', *Collected Essays,* Vol. 3 (London 1893).
[6] T. S. Moore and J. C. Phillip, *The Chemical Society, 1841–1941* (London 1947).
[7] P. J. Hartog, *The Owens' College* (Manchester 1900).
[8] *A History of the Cavendish Laboratory, 1871–1910* (London 1910).
[9] A. C. Chapman, *The Growth of the Profession of Chemistry, 1877–1927* (Institute of Chemistry, London 1927).
[10] R. B. Pilcher, *History of the Institute of Chemistry, 1877–1914* (Institute of Chemistry, London 1914).
[11] D. A. Winstanley, *Cambridge in the Eighteenth Century* (Cambridge University Press, 1922).
[12] S. P. Thompson, *Michael Faraday: His Life and Work* (London 1898).
[13] E. F. Armstrong, 'The Influence of the Prince Consort on Science', *Journal of the Royal Society of Arts,* Vol. 94 (23 November, 1945).
[14] *The Life and Work of Prof. W. H. Perkin, Jnr.,* (The Chemical Society, London 1932).
[15] Patrick Colquhoun, *A new and appropriate system of education for the labouring people* (London 1806).
[16] W. Richardson, *The chemical principles of the metallic arts* (London 1790, 2nd ed. 1806).
[17] Anon., *Observations on Mr. Brougham's Bill* (1821).

[*18*] 'The Eloquent Speeches of Dr. Birkbeck and Mr. Brougham', (at the opening of the New Lecture Room, Southampton Buildings, on 8 July, 1825).

[*19*] *Annual Reports of Mechanics' Institutes* (1825–   ).

[*20*] 'Paul Pry', *Blunders of a Big-Wig'* (1827).

[*21*] Rowland Detrosier, *The Necessity for an Extension of Moral and Political Instruction among the Working Class* (1831).

[*22*] *The Deed of Settlement of the London University* (11 February, 1826).

[*23*] *Letter to Shareholders and Council of the University of London* (U.C.L.) *on the Present State of that Institution* (1830).

[*24*] *Statement of the Proceedings towards the Establishment of King's College, London* (1830).

[*25*] *By-Laws for the Management and Discipline of the University of London* (1832).

[*26*] *Preliminary Statement of the Arrangements for Conducting the Various Departments of King's College, London* (July 1831).

[*27*] *Petition for Incorporation of the University of London* (1834).

[*28*] *Scheme of Studies for the Degrees Granted by the University of London* (1838).

[*29*] Olinthus Gregory, *Aids and Incentives to the Acquisition of Knowledge* (1838).

[*30*] J. W. Lubbock, *Remarks on a Classification of Knowledge* (1838).

[*31*] G. Dodd, *British manufactures—chemical* (London 1844).

[*32*] G. Dodd, *British manufactures—metals* (London, 1845).

[*33*] *Minutes of Committee of London University* (1837–40).

[*34*] *Rules and Orders of the* (London) *Mechanics' Institution* (1823).

[*35*] Rev. C. A. Row, *Letter to Sir Robert Inglis* (1850).

[*36*] J. Struthers, M.D., *How to Improve the Teaching in the Scottish Universities* (1859).

[*37*] *Abolition of the Tests at Oxford and Cambridge: Reports of Speeches at Manchester* (1866).

[*38*] W. J. Unwin, *Prussian Primary Education* (1857).

[*39*] Hely H. Almond, *Mr. Lowe's Educational Theories* (1868).

[*40*] W. G. Armstrong, *et al.*, *Industrial resources of the Tyne, Wear, and Tees* (British Association, London 1863).

[*41*] Correspondence relating to the first London University degrees.

[*42*] W. P. Atkinson, 'Classical and Scientific Studies in the Great Schools of England', read before the M.I.T. (1865).

[*43*] H. von Helmholtz, *On the Relation of the Natural Sciences to the Totality of the Sciences* (Heidelberg 1869).

[*44*] J. H. Bennett, *Physiology as a Branch of General Education* (1869).

[*45*] A. W. Williamson, 'A Plea for Pure Science', Lecture at University College, London (1870).

[*46*] P. L. Simmonds, *Science and Commerce* (London 1872).

[*47*] Charles Dickens, 'The English Peoples' University', *All the Year Round* (16 July, 1859).

[*48*] 'Degrees at London University', *Cassell's Family Magazine* (December 1875).   ,

[*49*] Swire Smith, *Educational Comparisons: or Remarks on Industrial Schools in England, Germany and Switzerland* (1873).

Index